C. C. LOCKWOOD'S
Louisiana Nature Guide

C. C. LOCKWOOD'S

Louisiana Nature Guide

Louisiana State University Press
Baton Rouge and London

OTHER BOOKS BY C. C. LOCKWOOD

Atchafalaya: America's Largest River Basin Swamp

The Gulf Coast: Where Land Meets Sea

Discovering Louisiana

The Yucatán Peninsula

First printing
04 03 02 01 00 99 98 97 96 95 5 4 3 2 1

Designer: Laura Roubique Gleason
Typeface: ITC Garamond
Printer and binder: Sung In Printing

Manufactured in the Republic of Korea

LIBRARY OF CONGRESS CATALOGING-IN-PUBLICATION DATA

Lockwood, C. C., date.
[Louisiana nature guide]
C. C. Lockwood's Louisiana nature guide.
p. cm.
Includes index.
ISBN 0-8071-1989-X (c : alk. paper)
1. Natural history—Louisiana—Juvenile literature. 2. Nature study—Louisiana—Juvenile literature. [1. Natural history—Louisiana. 2. Nature study.] I. Title. II. Title: Louisiana nature guide.
QH104.5.L8L63 1995
508.763—dc20 94-38613
CIP
AC

To Adrienne, Emily, Janey, Jennifer, and the Roland kids

Nature does not like to be anticipated . . . but loves to surprise; in fact seems to justify itself to man in that way, restoring his youth to him each time—the true fountain of youth.

—Walter Inglis Anderson
May, 1959

Contents

Foreword

After twenty-six years in conservation education, I take its principal question to be, How do we inspire curiosity about the natural world in our children? A combination of encouragement, information, and firsthand acquaintance seems necessary. It is no accident that the standard of nature-center education is the interpretative trail walk, for it approximates the experiences through which the marvels of nature captivated the naturalists themselves.

When I think about how I came to my own lifelong passion for the natural environment, I recall habitats, people, books, and . . . snakes. The fascination began in my early school years on trips from southern Alabama to our hometown in central Texas that followed U.S. Route 90 through New Orleans, Houma, Franklin, Lafayette, and Lake Charles. As a young child looking out the car window, I was sure life in south Louisiana meant having a boat and that each day brought encounters with alligators, egrets, and other awesome swamp and bayou critters.

In 1958, when my family moved to the banks of Rapides Bayou just west of Alexandria, I became familiar with the archaic garfish that swam right under the bayou's surface, as well as with the green tree frogs that caught insects on our front porch, the lubber grasshoppers that clambered about under the herbs, the mysterious snakes that slithered along the banks, and the incredible aromas that rose out of the humid forests truly encircling our house. Later, living in the east end of town, I came to know Sandy Bayou and the Red River and its bar pits. On weekends the family made outings to Camp Livingston, near Pollock, where we found sandy-bottomed, crystal-clear creeks full of madtoms, colorful darters, tadpoles, and salamanders. Habitats like the ones I explored can pique a child's curiosity.

My father and I took afternoon walks, maybe peering together at some strange creature under a log. The walks were stimulating, and my father's involvement showed that he valued and was glad to encourage my interests. It was also lucky for me that my mother tolerated a natural history museum in my bedroom, with all its snakes, turtles, and—occasionally—spiders. A local businessman, Paul Adams, gave me opportunities that turned me into the youthful curator of a small collection of snakes at a small community zoo. Doris Cochran, the curator of amphibians and reptiles at the National Museum of Natural History, tirelessly answered a neophyte's questions. Many a friend spent hour upon hour at my side awaiting just one more natural encounter, and the palpable engrossment drove my inquisitiveness to a deeper level.

Of the many books that propelled my interest, two stand out: Karl Kauffield's *Snakes and Snake Hunting* made my toes tingle in anticipation of the next field trip, and Raymond Ditmars' *Thrills of a Naturalist's Quest* steered me toward my professional specialization in tropical snakes.

As to the snakes themselves, I was smitten from the start. In my hunts for them, I came in touch with the diversity of nature and learned the insistence of the questions, Why? and, What's this?

C. C. Lockwood's Louisiana Nature Guide will play a large part in building our children's curiosity about the natural world. C. C. and I had to educate ourselves concerning Louisiana by reading books that dealt with other geographic areas, but today's children will have this excellent volume on our state's flora, fauna, natural features, and environmental situation.

This book is a solid introduction to Louisiana's natural history, written in the folksy vernacular that is C. C.'s trademark and illustrated by his sensational photographs. The visual power of the volume makes young readers want to keep turning the pages and returning to them, and the information, disarmingly conveyed as it is, amounts to a sound scientific basis that will stay with them for life.

Of as much importance as the scientific content is the practical guidance toward active participation in nature that C. C. offers in sections like those entitled "Backyard Wildlife," "Jobs in Nature," "Photographing Nature," "How to Take Your Parents Camping," and "Nature Places in Louisiana." The book not only stimulates interest but helps school-agers function as naturalists.

C. C. Lockwood's Louisiana Nature Guide is a welcome addition to my library, and I will make certain that all the youngsters in my family have copies. It provides the framework for an enduring enjoyment of the bountiful natural resources of the state and points the way toward their safekeeping.

ROBERT A. THOMAS

Louisiana Nature and Science Center, and
The Audubon Institute, New Orleans

Acknowledgments

First I would like to thank the thousands of kids who participated in the field trips and research for this project. Many appear in these pages, but the many who don't are just as important. Each and every kid, parent, and teacher I met has had a special part in the making of this book.

A few institutions went above and beyond in helping me, and I list them here: the Audubon Institute, especially the Audubon Park and Zoological Gardens and the Aquarium of the Americas; the Greater Baton Rouge Zoo; the Louisiana Nature and Science Center; the Louisiana Wildlife and Fisheries Commission, and its Natural Heritage Program; the Louisiana Wildlife Federation; the Nature Conservancy of Louisiana, especially its outings program; the Museum of Natural Science, as well as the Department of Geology and the Department of Zoology, at Louisiana State University; and The Backpacker store. Please forgive me for the names I accidentally leave out.

I also wish to thank the people at Louisiana State University Press for their help in making this a delightful project and a successful book. Les Phillabaum, Cathy Fry, Margaret Dalrymple, Laura Gleason, Claudette Price, and Barry Blose had a hand in the conception, design, or editing of the book.

C. C. LOCKWOOD'S
Louisiana Nature Guide

Introduction

When I grew up, nature was my picture show . . . my playground. It turned out to be quite educational, and I would not trade my childhood for anything in the world. Back then, store-bought recreation was unusual, and of course, I didn't have a computer or video games. On television there was only one channel.

Luckily, though, I lived at the edge of a pretty big town. Woods were nearby, which everybody called the brick plant woods. This wonderful afternoon hideout was a mile long, with hills about a hundred feet high on each side. There was a tiny creek at the bottom of the valley. Often the creek almost dried up during the summer, but a few small water holes always protected some bluegill, minnows, and crawfish. Over one of the hills, the massive machines of the brick plant dug into the hillside for clay, making a great hollow and two pinnacles, which we called the Grand Canyon, Big Goatie, and Little Goatie.

My friends and I spent hundreds of days climbing those goaties, hiking the woods, picking berries, and watching the birds, squirrels, raccoons, and even a pair of red foxes. We used to catch tiny crawfish, put them on a small hook, and snag bluegill, which we then cooked on our boy-scout servo stove.

In our backyards we caught toads on spring nights, and fireflies on summer nights. In the fall, we might try to talk some new kid into holding a bag open and calling for snipes. We were all tricked into doing it once. On the other side of the hill I lived on was Bloody Cave. You could walk in only about ten feet, but it had an opening about six inches square in the back. When we peered in with a flashlight, we could see no end. That bottomless hole must have been the start of the Legend of the Monster of Bloody Cave. During the second grade, I spent my first night alone in the tree house and was terrified for most of the night, until I made friends with the Monster.

Just a little way out of town we had the Ozark Mountains, Clear Creek, Hurricane Creek, and the Mulberry River. Mom and Dad took us camping a lot. Depending on the weather, we slept on the ground or in the tent. So that we could play in the creeks, Dad made us a simple canoe out of a rippled sheet of tin, tarring a piece of wood to each end. He made one that was six feet long when we were seven and another that was ten long when we got older. They sank like rocks if you turned them over—which happened a lot. When we were younger, Dad had to swim out and grab a red-and-white plastic fishing cork on the end of a fifteen-foot rope and retrieve it. Later he made us do it. All the creeks had rapids and were full of fish, frogs, and snakes. My dad was never much of a fisherman or hunter, even though he taught me to fish and hunt, but he was a frog gigger. Some of my fondest memories are of wading up Hurricane Creek behind Dad. I had to stay close, because the carbide lamp strapped to his head was our only light. As soon as he caught enough big old bullfrogs, we knew we'd have eggs with frog legs for breakfast. What a breakfast. I can smell it now.

We bodysurfed the rapids in life jackets,

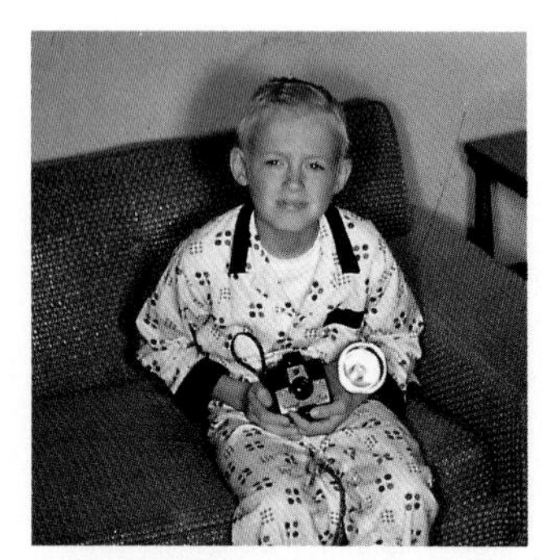
C. C. Lockwood holds his first camera.

and we once tried to sneak up on a great blue heron fishing along the creek bank. We hiked on the ledges of the cliff and watched swallows making their nests under the overhangs. Some of the greatest fun I had growing up was on these outings.

It's easy to think that kids—I mean younger kids, since I'm still a kid at heart—don't have the opportunities for outdoor exploring that I had. Today there are bigger cities, and highways, cars, noise, and posted land are everywhere. People seem to be straying farther and farther away from nature. But then something happens—the way it did while I was working on this introduction—to show how close at hand nature still is. An unusual bird flies by my window, not in ordinary flight but dodging acrobatically through the spring-fresh foliage of walnut, hackberry, and sweet gum trees. It is a wood duck, the multicolored male, and he lands on a limb twenty-five feet high in a sweet gum, near two nest boxes I have on that tree. The female is inside one of them, sitting on a clutch of thirteen white eggs, each just a bit smaller than a chicken egg. It's her second clutch. Her first ducklings bailed out of the box last month, waddled over to the drainage ditch, and swam to the lake two blocks away.

I also see a common grackle chasing one of the four gray squirrels I have in view at the moment. A cardinal sits on her nest in a sweet olive shrub twenty feet from my chair, as the colorful male calls from a tree on the other side of the deck. I can see a tufted titmouse, a Carolina chickadee, a starling, and a blue jay perched in nearby trees, as a brace of chimney swifts circle overhead, just below a solitary Mississippi kite high in the sky. I can hear a mourning dove, a yellow-billed cuckoo, and a bullfrog. Last night I saw a raccoon and an opossum foraging in my yard. Pretty good for the city. And Baton Rouge is four times as big as the town of Fort Smith, Arkansas, where I collected my childhood memories.

I can identify ninety-seven species of birds that have passed through my yard in the years I have watched it. Many nest there. Mammals, reptiles, and insects use my yard, too.

Within two hours, I can get to swamps, marshes, bays, bogs, bayous, creeks, lakes, hills, waterfalls, or piny woods. If I really wanted to work at it, I could see a hundred different species of birds in a weekend. I could hike, hunt, fish, swim, paddle, and camp. I could look for fossils, mushrooms, wild berries, and flowers. If I wanted to experience nature while keeping my shoes clean, I could see wildlife and plants in museums, nature centers, galleries, zoos, and aquariums, or I could attend lectures.

I fool myself if I think nothing like the opportunities I had when I was young are here now. There may be some differences, but out there in the great wild blue yonder, there is something for each and every kid to enjoy. We just need to make the effort to do it.

Why learn about nature? Because it's fun to be outdoors observing bears, bayous, and buttonbushes. We need to know more about the world we live in and about the other creatures in it. That way, we'll learn more about ourselves and how to survive and thrive in a rapidly growing and changing world.

Be a weekend naturalist and explore Louisiana's outdoors by keeping a journal,

The star anise has a different smell from other flowers. Some people say its odor is that of baby diapers. P . . . ew!

The raccoon, with its masked face, looks like a bandit.

Some dendrologists, who study trees, think bald-cypress knees give the trees support in soggy soil, but others believe they are a breathing mechanism for the flooded roots.

This lubber grasshopper is commonly known as the devil's horse because the underside of its wings is red.

A mycologist would call this mushroom *Amarita unbonata.*

Three young roseate spoonbills eagerly wait for a parent to regurgitate their breakfast.

taking a photograph, or drawing a sketch of what you observe.

More specifically, you could be a

Zoologist, and study all kinds of animals, including worms, snails, crawfish, spiders, fish, reptiles, amphibians, birds, and mammals—not leaving out us, *Homo sapiens*

Mammalogist, and study mammals

Ornithologist, and study birds

Herpetologist, and study reptiles and amphibians

Ichthyologist, and study fish

Entomologist, and study insects

Botanist, and study algae, lichens, flowers, trees, and shrubs

The cinnamon fern comes up coiled and unwinds as it grows.

Ecologist, and study the relationship of plants and animals to their environment

Environmentalist, and work toward protecting the environment from pollution and other harms

Geologist, and study what the earth was like in ancient times

Archaeologist, and study and recover evidence of human life and culture in the past

Paleontologist, and study fossils and ancient life forms

Astronomer, and study the sun, the moon, the planets, and the stars

Pick your favorite ology and learn more about it.

A group of kids learn a bit of astronomy by the campfire as the earth's rotation makes the stars appear to leave trails.

The Naming Game

Everybody loves animals! But when we say the word "animal," most of us think of a mammal, like a majestic white-tailed deer buck or a cunning cougar. Yet there are many other creatures in the animal kingdom: birds, reptiles, amphibians, fish, insects, crustaceans, flatworms, to mention only some.

Most visible living things that aren't animals are plants: trees, flowers, shrubs, and herbs, for example. But in rare cases living things cause puzzles for someone who tries to decide whether they're animals or plants.

Would you say that what's in the picture below is a plant or an animal? It's a euglena, which scientists consider at the boundary between plants and animals. It has a chlorophyll-producing structure, like most plants, yet it moves in an animal-like way and absorbs food when necessary.

A euglena

Plants are defined as green organisms that have no power of locomotion, are self-feeding through photosynthesis, and have cells with walls. Animals are defined as organisms that are capable of locomotion and of capturing their food and that have cells with no walls.

To sort out the millions of plants and animals around the world, we needed a better system. Carolus Linnaeus, a Swedish naturalist, devised such a method for classification, or a taxonomy, that divides living things into the plant, animal, and three other kingdoms. It also gives each kind of plant and animal its own two-part name (binomial nomenclature). To have the same two-part name is to be of the same species. For two animals to be of the same species, they must be able to breed and pass to their offspring their own characteristics. Thus the mule deer of the western United States is given a slightly different scientific name from our white-tailed deer. The mule deer is called *Odocoileus*

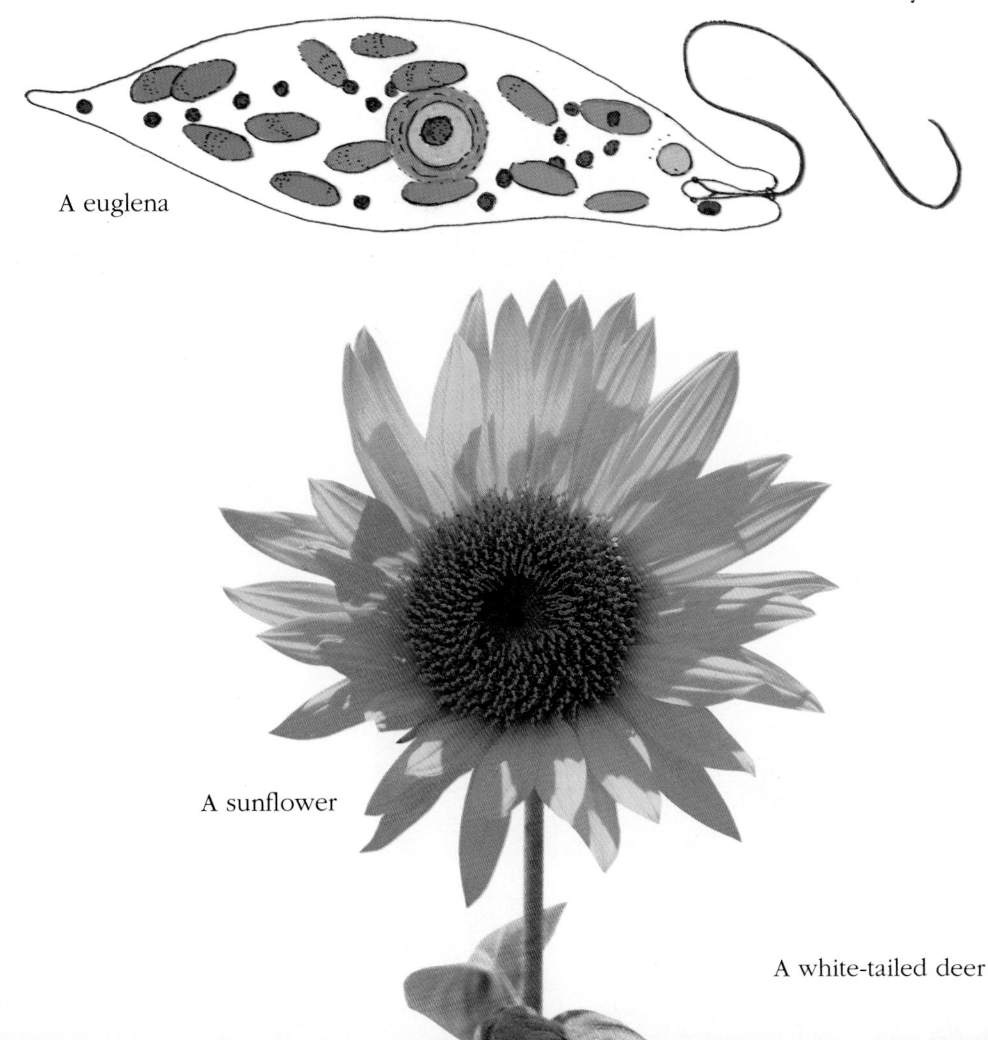

A sunflower

A white-tailed deer

The cougar, very rare in Louisiana, has numerous common names, including "mountain lion," "panther," "puma," and "catamount." By using the scientific name *Felis concolor,* zoologists eliminate all confusion.

hemionus, and the white-tailed deer, which lives in Louisiana, is called *Odocoileus virginianus*. The sunflower's name in taxonomy is *Helianthus annuus*. Look at the chart below to see how each is named from kingdom to species.

If you saw a white-tailed deer, you'd probably refer to it as "the deer." In Mexico, a student your age would call it *el venado,* which in Spanish means "the deer." But a Mexican scientist would call it *Odocoileus virginianus,* just as an American scientist would. Zoologists from different countries need a way of making sure they're talking about exactly the same thing.

The three kingdoms in addition to Animalia and Plantae are Monera (single-celled organisms, such as bacteria), Protista (organisms like algae), and Mycota (fungi, such as mushrooms).

To learn more about the naming and classification of plants and animals, check out a botany or zoology textbook from your local library.

WHITE-TAILED DEER

Classifications		**Sample Characteristics**
Kingdom	Animalia	Moves, eats
Phylum	Chordata	Has nerve cord
Class	Mammalia	Has mammary glands
Order	Artiodactyla	Has even-toed hooves
Family	Cervidae	Is related to elk, moose
Genus	*Odocoileus*	
Species	*virginianus*	

A young alligator

SUNFLOWER

Classifications		**Sample Characteristics**
Kingdom	Plantae	Contains chlorophyll
Division	Anthophyta	Forms flowers
Class	Dicotyledonae	Has two seed leaves
Order	Asterales	Has many-petaled flowers
Family	Asteraceae	Is related to coneflowers, asters
Genus	*Helianthus*	
Species	*annuus*	

Bones of Louisiana's Past

Imagine tall spruce and fir trees lining the banks of Bayou Sara as thick fog rolls up the valleys. Hiding behind one big tree trunk are three nomadic Indian hunters with flint-tipped spears. Up the bayou, a four-ton mastodon, an elephantlike animal, moves toward some spruce branches to browse. More Indians are hiding farther upstream, waiting for their friends to thrust the first spear in. You can guess what happens next as one Indian runs in front of the beast to get its attention. While he's being knocked nine feet into the air by the great tusk, the other two lunge their spears into the mastodon's rib cage. Soon after, there is a raging fire and others come to feed at a big barbecue.

All this took place about eleven thousand years ago, during the last Ice Age. Geologists call it the Pleistocene epoch. The sea level was 450 feet lower than now, because a lot of ocean water was frozen in the polar ice cap. Louisiana's coast was fifty miles farther south in some places, since the entire continental shelf was out of the water. Along with the mastodon, there were camels, horses, mammoths, and giant ground sloths.

Two brothers, Mark and Chris LoCoco, look at fossils and a tooth found on an outing to Como Creek conducted by the Nature Conservancy of Louisiana. Creek beds are a great place to search for fossils.

The triceratops did not roam Louisiana, but because of the Audubon Park and Zoological Gardens, in New Orleans, Iam Tucker and Danny O'Neill can see what that reptile of 200 million years ago was like.

Then a climate change occurred, making it warmer. The ice began to melt. The sea level was rising. The big animals probably moved north, with the Indians hot on their trail. These animals were slow breeding, slow moving, and probably pretty stupid. Scientists think the Indians killed them off. Four thousand years later, buffalo were what were left for the Indians to hunt. The buffalo bred fast enough to take the hunting pressure from the Indians, but later they were no match for the modern human's gun.

Long before the Indians were hunting, during the time that dinosaurs roamed the world, Louisiana was arid—much like Utah today. That was during the Jurassic period, 200 million years ago. Late in the Jurassic, the oceans began to cover Louisiana, and by the Cretaceous period, 144 million years ago, North America was divided by an inland sea. Louisiana was totally under water. Giant sea turtles swam over Louisiana, as did a thirty-two-foot marine reptile classified as *Elasmosaurus,* which I picture as looking like a fantasy creature, maybe like the Loch Ness monster. It was a shallow sea, and many other creatures lived and died there. Some left fossils.

The early natural history of Louisiana is told by fossils. Geologists, archaeologists, and paleontologists find these records and arrive at conclusions about how Louisiana looked. Fossils of sharks, whales, barracuda, sawfish, and eagle rays have been found in Bossier Parish. Bones of mastodons and giant sloths have been found in East Feliciana Parish. Elsewhere in Louisiana, fossils of mammoths, crocodiles, oysters, clams, corals, and crinoids have come to light. Most have been discovered on land, but shrimpers dragging their nets across the continental shelf have trawled up mastodon teeth.

Louisiana has changed much during the millennia. The weather and the land forms will continue to change over the eons. In the short term, though, we must take care not to be the cause of disasters. Nature destroys habitats and makes animals go extinct, but human beings shouldn't.

Look for fossils in creek beds. When you find one, try to imagine what the earth looked like at the time that the plant or animal died. The Tunica Hills are a great place for this.

Alton Dooley, a graduate student at Louisiana State University, shows some kids the jaw and tooth of a four-ton mammoth found in West Feliciana Parish.

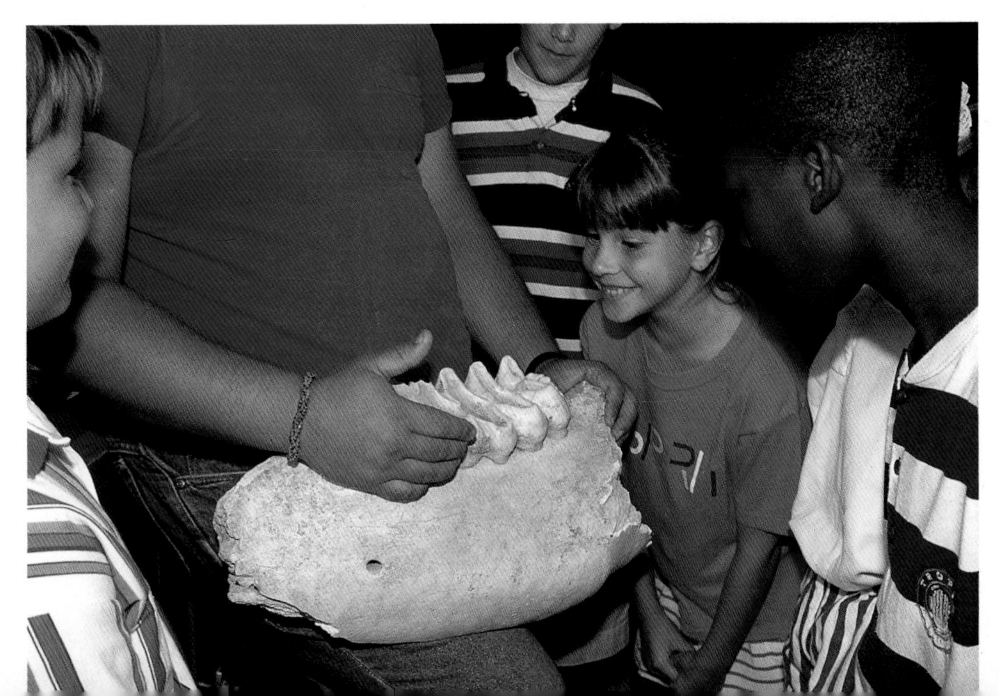

Mammals

NUMBER OF SPECIES

Worldwide	4,000
United States	470
Louisiana	70

A mother raccoon nurses her quadruplets.

A white-tailed buck

We are mammals, and we share a few characteristics with chipmunks, deer, nutrias, pigs, and whales, and with everything else in the class of Mammalia. These characteristics make us different from and more advanced than some other animal life. Two of the traits belonging to mammals are hair (fur) and mammary glands. Female mammals have mammary glands so they can feed milk to their offspring. In spite of being fully developed at birth, young mammals need parental care—from as little as a few weeks, in the case of mice, to many years, in the case of human beings. Mice and men are related, however strange that seems.

It seems just as strange that the southeastern shrew, Louisiana's smallest mammal, is related to our largest, the blue whale. It would take sixty-two million shrews, each weighing about as much as a penny, to equal the weight of a single 196-ton blue whale. To see these two critters, you would have to be very lucky. The blue whale is rare in Gulf waters. Dolphins are the most

An eastern chipmunk

tail shrew shown here has a venomous bite and can kill small mice.

Our biggest terrestrial animal is the Louisiana black bear. A male bear can weigh up to six hundred pounds, but it is very hard to get close enough to see it well in the wild. I sat in a blind one day and watched a bear pass thirty feet in front of me. The big creature caused no sound, and even with its lack

A short-tail shrew

common marine mammals in Louisiana, and you can see them regularly around Grand Isle and in other coastal waters.

The three species of Louisiana shrews hide in the leaf litter of forests. Turning over rotten logs is about the only way to see them. They are voracious animals, usually eating their weight in insects each day. The short-

A dolphin swims in the Gulf of Mexico.

The opossum is Louisiana's most primitive mammal.

Crepuscular, white-tailed does and fawns feed at dusk.

of camouflage, it was almost invisible in the green woods. Most mammals are hard to see because they are secretive, good at camouflage, and nocturnal (active only at night) or crepuscular (active at dusk and dawn).

Mammals that are diurnal (active in the day), like the chipmunk, are easier to observe. In Louisiana, chipmunks are found only north of Baton Rouge, in the Tunica Hills. Look for them scampering among the low branches and forest floor in search of nuts and seeds. The most primitive mammal in Louisiana is the opossum. It raises its premature babies in a marsupial pouch until they're developed enough to crawl out on the mother's back. I see opossums in my muscadines every summer. These southern grapes attract much wildlife to my yard.

White-tailed deer can be seen in the evenings as they come out of cover to drink and to feed in fields. They are crepuscular.

A playful thing to do with a squirrel is to tie a pinecone on a foot-long string and coat it with peanut butter and sunflower seeds. Hang it from a tree limb, and see how long it takes the squirrel to figure out how to eat it.

Birds

The great egret grows fancy nuptial plumage for each nesting season.

NUMBER OF SPECIES

Worldwide	9,000
United States	650
Louisiana	430

Fly through the air! What a neat deal. Insects do it, bats do it, but we most enjoy seeing birds do it. How do they manage it? Birds have a hollow, lightweight skeleton and a big breastbone, called a keel, to which strong muscles are attached. These muscles move the feathered wings. It's not really as simple as it sounds. It took millions of years for birds to evolve the way they did. What does flight give them? A means of escape, yes. A means of locomotion, yes. A means of finding and securing food, yes.

Feathers are a characteristic that only birds have. Some feathers are for flight. Others, like the down of baby birds, are for warmth. Still others are for decoration. Dress-up feathers have sometimes endangered the birds that have them. The breeding plumage of the great egrets is so beautiful that it almost caused the birds' extinction when market hunters slaughtered them in their rookeries in order to get feathers to sell to the makers of fancy women's hats. The Audubon Society was founded to stop that. In 1909, it helped get the Lacey Act passed, banning trade in egret plumes. Since then, the egrets have recovered.

The osprey is sometimes called the fish eagle.

All birds molt. In other words, they lose and replace their feathers. Some molt twice a year and show a more colorful set of feathers during the breeding season. The male wood duck is beautiful year round, but it is especially beautiful in January and February.

The sizes and shapes of feathers vary between kinds of birds. Some feathers have evolved for cold climates, some for fast

Long legs help this black-necked stilt feed in the mud of the marsh.

A turkey vulture suns its wings.

flight, and some for use in water. The anhinga can swim under water, and its feathers have no oil coating to keep them dry. This strange bird has to dry its wings after it swims. Although the turkey vulture looks as if it is doing the same thing, it isn't. The turkey vulture suns its wings to get rid of parasites.

All the kinds of birds in Louisiana make good parents except for the cowbird, which lays its eggs in another species' nest and lets the nesting bird raise its young. Ornithologists think the cowbird evolved this pattern by following the nomadic buffalo herds and never being in one place long enough to build a nest and hatch its eggs.

An anhinga dries its wings.

Chicks are helpless and depend on their parents for food, shelter, and protection. Most young birds look pretty silly until they fledge. Fledging is acquiring the feathers that will let them fly away from the nest. Even after they can fly, though, most will hang

The feathers of the cute but homely little blue heron are white until it reaches a year of age, when it molts and grows blue plumage.

This prothonotary warbler, like most birds, has to feed its helpless chicks.

around their parents to beg for food. One- and two-year-old male red-cockaded woodpeckers will stay in a colony with their parents and help feed their new brothers and sisters.

Since Louisiana is on the Mississippi flyway, a corridor of migration, and since Louisiana's habitats are so diverse, we have more species of birds than most states. Only five states—Arizona, California, Colorado, Florida and Texas—have more than we do, and those states are all bigger than we are.

You can see Louisiana's smallest bird from the spring through the fall. It's the 3 1/2-inch ruby-throated hummingbird. In the winter, look for the white pelican, whose wingspan is more than eight feet. It has the largest wings of any bird in the United States.

Bird watching is a popular hobby. Some birds, like the wild turkey, are easy to identify, but others, like the sparrows and warblers, are much harder. Bird guides list field marks to help you distinguish between similar birds. Knowing the field marks—perhaps wing bars or a colored throat patch—makes identification much easier. Start a checklist, and see how many Louisiana birds you can chalk up.

The turkey is Louisiana's heaviest bird.

Reptiles

NUMBER OF SPECIES	
Worldwide	6,000
United States	283
Louisiana	96

Like most reptiles, alligators hatch from eggs.

Reptiles are more primitive than birds and mammals, and they evolved much earlier. The first reptiles were on the earth 315 million years ago. Most reptiles lay eggs, as birds do, but they have scales instead of feathers and are cold-blooded: their body temperature is the same as their surroundings. Most reptiles have to burrow down somewhere to avoid extreme heat and cold.

Reptiles are called precocious, meaning that they can fend for themselves as soon as they are born. This alligator hatching from its egg can feed itself, but young alligators will stay near their mother for a while for some protection. These one-day-old Texas rat snakes may never see their parents. The mother snake laid her eggs in a pile of leaves and left.

Most snakes smell with their tongues.

One-day-old Texas rat snakes

Snakes have highly-developed Jacobson's organs and can smell with their forked tongues. Rattlesnakes and other pit vipers have loreal pits below their eyes; these are sensitive enough to heat radiation that the snakes can locate mice by their warmth and strike them in the dark. Although snakes have normal-looking eyes, they don't see well.

Louisiana has forty species of snakes, and six of them are poisonous. You should use extreme care around snakes and alligators.

Sea turtles are well known these days. It's just too bad that their fame stems from being endangered. These gentle swimmers nest on beaches worldwide. They're in trouble because everybody likes to eat their eggs. Even people eat them. Another problem is caused by all the hotels and condominiums being built on the sea turtles' nesting beaches. Often there is too much light and activity on the beach for the mother turtles to nest. If

The pattern of the scales on the eastern diamondback rattlesnake gives the reptile its name.

she does lay her eggs, the confused hatchlings may head to the light of the condos instead of the sea.

Sun-loving freshwater turtles love to sit on logs to catch the rays, only to jump in with a splash when they're disturbed.

Lizards too are reptiles. The Mediterranean gecko is an exotic species that got to Louisiana by other than natural means. The near-transparent nocturnal lizards probably arrived on a cargo ship from the tropics. They must like it here, for they are reproducing well. I have twenty that climb all over my house. At night, they usually hang around my windows eating bugs attracted by the light inside.

The alligator snapping turtle held here by Marty Stouffer, the host of *Wild America,* has a bad reputation. When I was growing up, I kept hearing that if a snapping turtle bit you, it wouldn't let go until the sun went down. That isn't true. Still, anyone would want to keep clear of its intimidating mouth. But like most other animals, the snapping turtle isn't going to bite you unless you trouble it.

Alligators are easy to find in Louisiana from April to October, and in some parts of the state it's possible to see them year round. The marshland in south Louisiana is the best place for coming across them. Visit the nature trail at Sabine National Wildlife Refuge. It has some big alligators, as well as a lot of other wildlife to see. Many swamp tours in south Louisiana can show you alligators too.

Marty Stouffer, the host of television's *Wild America,* visits Louisiana to look for alligator snapping turtles.

Marshall Morgan shows Ariel Roland the rare gopher tortoise he has found (top). Marshall turned the reptile over to the Louisiana Department of Wildlife and Fisheries' endangered species specialist, Richard Martin, after he discovered how uncommon it is (bottom). Later it was released in its native habitat.

A Mediterranean gecko

Amphibians

NUMBER OF SPECIES

Worldwide	3,900
United States	194
Louisiana	49

Stephen Cropper, Jeff Merchant, and Macon Roland study two marbled salamanders before putting the amphibians back under the log where they were found.

The word "amphibian" means something that lives (bios) both (amphi) on land and in water. Long ago all of what are now amphibians lived in water, but eventually most of them developed lungs that allowed them to venture onto land. Amphibians have been around for 360 million years. Most have a hairless skin that must be kept moist or they will dry out. Among the amphibians, frogs and toads are the most common. Tadpoles live in water full time, and later the adult frogs live in and out of the water. There is also a diverse group of lizard-shaped amphibians that are secretive and slimy. These are the salamanders, including newts, water puppies, sirens, and congo eels. Salamanders are very hard critters to see in the wild.

Frogs are very vocal, and it is surprising how loud a group of spring peepers can sound after a rain. A summer camp-out is not complete without hearing the croak of

A green tree frog

A marbled salamander

takes a lot of work and special cages to keep most animals happy and healthy if they're removed from their natural habitat.

Get your parents to help you make backyard wildlife habitats. By doing that, you can observe wildlife in your yard without penning it up. Keep a journal of all the animals you see in your backyard.

the peepers and the muffled roar of a bullfrog.

A group of kids and I went salamander hunting with Dr. Bob Thomas, a herpetologist, in November. We turned over fallen logs with hooked sticks, because we didn't want to place our hands near a hiding snake. After four hours, we had observed three marbled salamanders, two slimy salamanders, three green tree frogs, two narrow-mouthed toads, and countless Bess beetles, spiders, centipedes, millipedes, and worms. Dr. Thomas told us always to put the logs back just the way they were. Otherwise, we'd destroy the animals' homes. Common reptiles and amphibians are easily caught and make good temporary pets. But remember, never disturb rare species, and after a couple of days of observing a box turtle or a green tree frog, return it to where you found it.

Wildlife should belong to no one; it should be there for everyone to enjoy. It

Who says all toads are ugly? Look at those beautiful golden eyes.

The pig frog is closely related to the bullfrog.

Fish

NUMBER OF SPECIES

Worldwide:	22,000
United States:	2,400
Louisiana:	450

The Age of Fish was 400 million years ago, when cartilaginous (boneless) fish, like sharks, and bony fish, like bass, were evolving. The fish developed a true jaw and appendages, which gave them advantages over the invertebrates living at the time. The appendages of one branch of the evolving creatures were lobed. That means they were fleshy—more like a leg than just a bony fin. These fish evolved into amphibians. The rest developed into red snapper, speckled trout, great white sharks, and the multitude of other varieties in the seas and streams.

Most fish we get to see are on the end of a fishing pole. The sac-à-lait, to use the Cajun name for the white crappie, is an especially tasty fish that anglers catch over brush piles or around bald-cypress stumps. Fishing is a great pastime, one of the most popular in the world. Whether you like to fish or not, it's fun getting out into nature. If you're not catching fish, you can observe other wildlife or be observed.

To see fish swimming, you must snorkel, scuba dive, or visit an aquarium. Some creeks are clear enough to let you gaze at a garfish, bluegill, or largemouth bass, but the visibility is much clearer in the Gulf of Mexico beyond the muddy water of the Mississippi. There you can see the ferocious-looking sand tiger shark, which in spite of its

The early bird gets the worm, and these two kids and their dads hope the early angler gets the fish.

Scuba divers photograph and hunt fish under offshore oil platforms (bottom). These structures draw many species, such as the sand tiger shark (opposite page, top), the colorful blenny, seen living in a sponge-encrusted barnacle (opposite page, upper middle), the cocoa damselfish, sometimes called the LSU fish by Tiger fans (opposite page, lower middle), and the lookdown, which forms schools and has a sad mouth and a flat silver body that make it seem the "goofy fish" (opposite page, bottom).

awesome teeth is the most docile of free-swimming sharks.

Fish often school—not to learn about math and history but to protect themselves from sharks and barracuda. When a predator charges the school, the mass confusion makes it hard for it to pick out one fish. Lookdowns are a common schooling fish in the Gulf. They can be seen by scuba diving under oil rigs.

Many marine invertebrates attach themselves to oil rigs' pilings—soft corals, anemones, shellfish, and barnacles, among them. This dead barnacle makes a home for a blenny, a colorful, clownlike one-inch fish.

The biggest fish in the waters in and near Louisiana is the whale shark, a harmless plankton filter feeder. It can reach fifty feet long. I saw one swimming off the tip of the Mississippi River that was larger than the thirty-three-foot boat I was in. It looks like a whale and feeds like a whale, but it's a fish. Scuba divers can swim with these gentle giants. They never bother people.

At the other end of the spectrum is the half-inch least killfish. This minnow lives in swamps and marshes and can survive in either fresh or brackish water.

Get a small aquarium and fill it with local fish. You can use a gallon pickle jar and a cheap bubbler from the discount store. You'll be surprised how many neat-looking minnows you can find in a roadside ditch. Just use a small dip net to catch a few. You'll also find many different species of invertebrates in your catch. Have a bucket of water with you to take them home in.

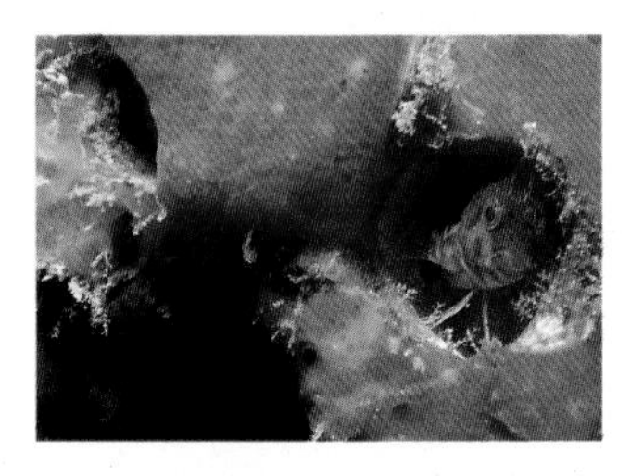

Invertebrates

A crawfish

There are myriads of other animals in Louisiana, including unicellular protozoans, worms, sponges, corals, starfish, shellfish, and the largest group of all, arthropods. Taxonomists have identified more than a million species of arthropods. Most of these are from the class Insecta, in which some entomologists think there are possibly thirty million species yet to be described.

Not only insects but also crawfish and spiders are arthropods. All have jointed appendages. It's easy to see the jointed legs on a crawfish, but you probably haven't seen the two limy pebbles there are in a soft-shell crawfish. Since these arthropods have their skeletons on the outside, they eventually run out of room to grow. Then they have to shed their shell. The crawfish retains its calcium in the form of pebbles, and later uses the pebbles to harden its larger exoskeleton.

Louisiana has twenty-nine species of crawfish, and all have a place in the dinners of

Fall webworms

The anemone is one of the many species of marine invertebrates.

Fire ants are able to live in flooded fields for months by taking turns on the top. The mass of ants rolls over and over, giving each insect time to breathe.

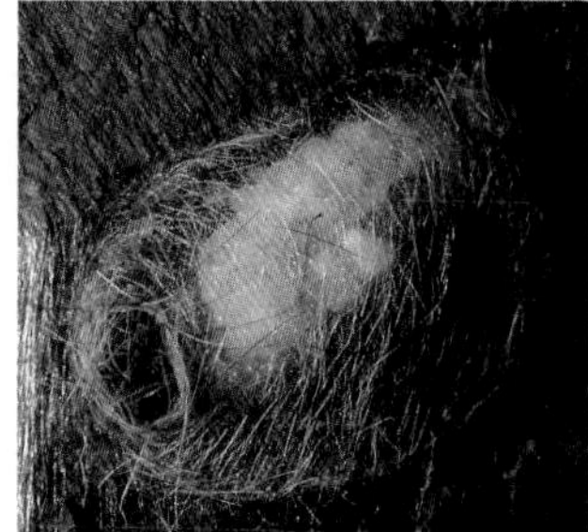

This future moth still has not reached its pupal stage in its cocoon.

birds, mammals, fish, amphibians, and reptiles. Only the swamp red and the river white varieties are suitable for crawfish boils, though. We catch about a hundred million pounds a year to put in our bellies. Other species that fancy crawfish catch many more pounds than that.

Insects make the crawfish numbers look paltry. Did you know there are a greater number of species of beetles than there are of species of plants in the world? There can be 75,000 fire ants in one mound, and seventy-five mounds on one acre of land. Fire ants are survivalists. If their mound is flooded in the spring, they can float in a ball. They take turns breathing by spinning around and around. These hard-stinging beasts are an exotic species that came from South America. They have become a major problem.

A six-spotted fishing spider

Think of a continuous line of inch-long fall webworms reaching up to the moon, circling it, and coming back. One female webworm would have that many great-grandchildren if all her offspring lived through one year. At any moment there are three hundred pounds of insects on the earth for every single pound of the human race.

What this means is that there are gobs of

A ghost crab

food for all the other animals on the earth. I have seen a cattle egret regurgitate a ball of twenty-five grasshoppers to its nestful of funny-looking babies. It takes a lot of little creatures to feed us bigger guys.

What else do insects do for us? Bees and butterflies pollinate our crops. Insects make honey, beeswax, and silk. They feed on and help decompose dead animals, and they turn dung and dead plants into usable nutrients. We also get medicine from insects.

Next time you slap a mosquito or step on a cockroach, think of what life would be like without insects. Without insects, most of the rest of the animal world wouldn't be here either. For without pollination, our corn, tomatoes, and other crops could not grow.

Take a magnifying glass and visit a vegetable garden. I'll bet you can find ten different insects in ten minutes if you look closely under the leaves, on the stems, and on the vegetables.

A monarch butterfly

The dread of every gardener is aphids on a tomato plant.

The Plant Kingdom

What would the world be without plants? To start with, there would be no potatoes on our plates. Of course, we would have no shade from the sun, no wood for our houses, and no flowers in our gardens. Without plants to produce oxygen, the rabbits, the deer, and we ourselves would not be here. Ninety-nine percent of all our oxygen comes from plants.

If you look at the earth as a structure of things supporting other things, you can think of it as a pyramid, with rocks, minerals, and mud at the base. The plants and lowest forms of animal life then make up the next tier of the structure. Its massive quantities of grasses, herbs, flowers, fruits, buds, and wood feed the tier above it, the herbivores. At the top are human beings and the other carnivorous animals, with the sun shining its energy down over everything. When we eat meat, a stupendous amount of sunlight, water, and plant life has gone into making each bite.

But this is a special kind of pyramid. If you take away any part, even if the part is near the top, the internal stresses will cause things to start to crumble throughout. We need all the pieces. If we lose a lower form of life, higher forms won't have what they need to survive. If we lose a higher form, it may cause an explosion of life in a pesty lower form, such as a mosquito or a noxious weed.

Plants feed us and the rest of the animal world. That's not all they do, for in absorbing sunlight, water, minerals, and carbon dioxide, they produce oxygen and retain water. Retaining water is more important than you might think. In the desert, many a prospector has survived on the moist pulp of a barrel cactus. More important to us in Louisiana are the thousand gallons of water a mature

NUMBER OF SPECIES	
Worldwide	248,000
Louisiana	3,000

The pyramid of life

The red maple, also known as the swamp maple, is one of the trees whose seed is called a samara. The winged seed flutters down and away from the tree like a helicopter.

The gracefully curving rattan vine is sometimes used to make furniture.

tree in the bottomland forest retains while its roots and leaf litter are holding the soil together. What happens when you cut 76 percent of the bottomland hardwood forest? That's what we've done in the alluvial plain of the Mississippi. The trees that held a thousand gallons of water have been replaced with fields, highways, houses, and parking lots. The result is faster rainwater runoff, less water storage, and therefore bigger and more frequent floods. Beyond making a home for wildlife, our forests prevent floods, produce oxygen, cleanse our water, and furnish lumber. Our forests are also a great place to hike.

The plant kingdom is as diverse as the animal kingdom. And like animals, plants do what is appropriate to continue their species. Pollen from a male flower must reach a female flower, and flowers have many tricks to get their pollen transported. The most common involves nectar, a sweet reward they offer the bats, birds, and bees that move pollen from one flower to another. Wind plays an important role in dispersing pollen, too.

An American lotus

An Indian pipe

After the flower is pollinated, a seed is produced. Some seeds just drop, but others are light enough to be caught in the wind. Still others develop within elaborate fleshy containers, or fruits, that attract an animal to eat them and leave the seeds elsewhere, often wrapped in manure. Some seeds have wings; others have sticky hair or fishhook spines to attach to an animal and be carried away. Some float and drift down a stream and reach the next sandbar. But there are also plants that grow vegetatively, without producing flowers and seeds. These just send out shoots that start a new plant. Each and every plant has evolved a system enabling its species to continue.

Among the strangest plants in Louisiana is the pitcher plant, which is carnivorous. A sweet substance attracts insects into its cave-like opening. Once they crawl lower in the tube, sticky downward-pointing hairs keep them from escaping. From the insects, the plant absorbs minerals it doesn't get from the soil. These plants have adapted to survive in poor soil.

The Indian pipe plant is an exception to the rules of the plant kingdom. It lacks the chlorophyll that gives plants their green appearance, but botanists still classify it as a plant. It's a saprophyte that lives off decaying organic matter.

In Louisiana's mild climate, some trees bear leaves and flowers even in the winter months. Our climate is part of the reason that we have a greater number of species of wildlife and more abundant wildlife than northern states. Diverse vegetation makes room for diverse wildlife.

From the 150-foot loblolly pine to the half-inch mosquito fern, Louisiana's plant life offers something to appeal to everyone.

Start a leaf collection. Pick one leaf from each of the different kinds of tree in your neighborhood. Dry and press your leaves between newspaper with a stack of heavy books on top. Better yet, make a press with two pieces of plywood about two feet square. Drill a hole in each corner. Use four 3-inch bolts with wing nuts to hold the two pieces together. Put newspaper between them. You can do about twenty leaves at a time. In about three days they will be flat and dry. Glue them to a sheet of white paper, and write beside each leaf the name of the tree it came from.

A pitcher plant

Habitats

A blue crab and shrimp

A colony of sandwich terns and royal terns nests on the Chandeleur Islands (bottom). The terns feed on the same varieties of shrimp and crabs that supply our tables from Barataria Bay (top right). The gulf habitats are where we find our seafood.

An animal's habitat can be defined as the place where it lives and feeds. Most animals can live only in the specific habitat they have evolved for. Loss of specific habitats is one of the main reasons certain species are endangered today. Humans are different from most of their fellow animals in being able to adapt to most of the conditions found worldwide. They are able to do this because they can take food, clothing, and shelter with them.

In a lot of states, the amount of rainfall is the cause of the greatest differences between habitats, but the rainfall doesn't vary greatly from one part of Louisiana to another. In our state, elevation and soil have a greater effect. We can divide Louisiana into five general kinds of habitat: gulf, marsh, swamp, prairie, and forest. But there are enormous differences within each of these.

The gulf habitats include the bays, the bar-

Louisiana's marshes take many forms, from rivers of grass among meandering bayous (top) to a sea of banana lilies that harbors countless species of wildlife (bottom right). Young nutrias can live among the banana lilies (bottom left).

rier islands, and the shallow waters of the continental shelf. Porpoises, fish, seabirds, and many invertebrates, such as shrimp, make this prolific area their home.

Marshes usually retain water and have nonwoody vegetation. They can be wet all year. Depending upon how much salt is in the water, they're described as salt, brackish, intermediate, or fresh marshes. Louisiana has 41 percent of the nation's coastal marshlands. Alligators, nutrias, shrimp, and many birds thrive in marsh habitats.

Swamps are wet forested areas. True swamps are wet most of the year; hardwood bottomlands are wet only part of the year. Bottomland forests are becoming rare now that we've clear-cut over half the trees in this habitat. All Louisiana's rivers have swamps

The swamps of the state range from pure stands of bald cypress (bottom right) to mixed bottomlands with a fairylike ground cover of ferns (top). Barred owls swoop down upon the swamps for crawfish and other aquatic edibles (bottom left).

by them. Crawfish, black bears, and barred owls like the wet woods of swamp habitats.

Prairies are extensive areas of dry land with grasses and very few trees. Louisiana's great prairie once occupied the triangle of land that has Lake Charles, Ville Platte, and New Iberia close to its corners. It is almost totally gone today, with farms covering the whole area. Whooping cranes and Attwaters prairie chickens once flourished in the prairie habitats of Louisiana. Neither is any longer found in the state.

Forests cover most of central, north, and eastern Louisiana. Around Covington and De Ridder there are lowland pine forests known as flatwoods. In Kisatchie National Forest, most of the area is upland pine forest. North Louisiana has mixed pine, oak, and hickory

forest. In the Tunica Hills lies an eastern deciduous forest of beech, magnolia, and oak. Deer, turkeys, and woodpeckers are some of the animals in Louisiana's forest habitats.

A fringe area where two general kinds of habitat intermingle—an ecotone—is usually very productive, ideal for animals with an ability to feed on both sides. The river otter, for instance, can fish in both the swamp and the marsh.

Louisiana is famous for its seafood, both for the large commercial catch and for the way we cook it. We have so much to harvest because the gulf, marsh, and swamp habitats blend together.

The interlocking of general kinds of habitat is what allows many invertebrates—notably shrimp, crabs, and oysters—to survive from one stage of development to another. The brown shrimp spawns in the Gulf in waters 150 to 360 feet deep. The fe-

The forests of Louisiana can be pure stands of pine, including giant loblollies like one specimen near Alexandria (bottom right), or hardwoods like those on a bluff in the Tunica Hills (top). Fungi such as mushrooms cover the forest floor (bottom left). Deer, squirrels, and turkeys hide among the trees.

There is no photograph of the prairie because the prairie in Louisiana has disappeared, along with the wildlife that occupied it.

Some species, like the river otter (middle), can live in widely diverse habitats, but others, like the red-cockaded woodpecker (bottom), have very specific habitat requirements.

male releases millions of eggs, which hatch and blend with the pasture of the sea, plankton. Much of the plankton is consumed by other animals, but the survivors from the hatching molt through eleven stages to become larval shrimp. These small critters make their way through islands and bays to grow up in the aquatic grasses of the marshland before going back to sea as adults.

Shrimp need the marsh, the bays, and the Gulf. In fact, 90 percent of our commercially caught fish and shellfish require more than one kind of habitat in their life cycle.

Remember, the loss of habitats is one of the principal reasons there are endangered species today. Each animal has to have a habitat.

Most of us live in an urban habitat. Figure out what the nearest natural habitat is to your home. Visit it often, and keep a checklist of the plants and animals you discover there.

The seasons occur because of three states of affairs: The earth makes a complete rotation on its axis every twenty-four hours. The earth orbits around the sun about every 365 days. The earth is tilted on its axis.

During the summer, the Northern Hemisphere is tilted toward the sun. Thus, as the earth rotates, we north of the equator receive more sun. With more sun, that part of the earth is warmer: the days are longer and the sun is at a higher angle. On or close to September 21 and March 21, we have the autumnal and vernal equinoxes, with day and night equally long. Only on those days do the two poles receive the same amount of sunlight. Then, in the winter, the Northern Hemisphere is tilted away from the sun, and it receives less sunshine and at a lower angle.

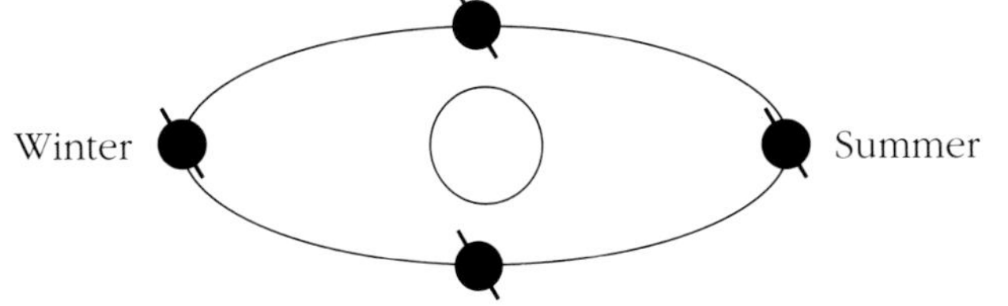

It's not just because of daylight saving time that we have long evenings. In Louisiana, we have fourteen hours of sunlight on June 21; that's four more hours than we have on December 21, our shortest day. North of

The Milky Way lights up a summer sky.

Water sports like conoeing are perfect summer pastimes.

The gray squirrel's world is different in summer, fall, winter, and spring.

Over a cornfield in Madison Parish, the harvest moon signals summer's end.

the Arctic Circle, there are twenty-four hours of daylight on June 21, and twenty-four hours of darkness on December 21.

With longer and hotter days, summertime is for swimming, canoeing, and water sports. Most animals are harder to see, because the foliage is thick. The gray squirrel doesn't have to feed all day long. At night you can see the Milky Way, our galaxy, in its full sparkling splendor. Lie on your back and watch for satellites.

As the days grow shorter and fall is in the air, it's time for hiking. The woods are cooler, and north Louisiana is painted in dazzling colors. Sometimes we get good fall colors in south Louisiana too, but it depends on the weather. The gray squirrel is eating acorns as the leaves change behind it. The

Fall is a great time for hiking, since the mosquitoes thin out and the leaves put on their prettiest dress.

harvest moon is the nighttime delight of this season. It gets its name because the extra hours of brightness it gives let farmers take in their crops in time.

Winter may bring some snow to north Louisiana and a lot of cloudy, cold, rainy days to south Louisiana. But what about those sparkly blue skies when a cold front comes through? Those are some of the best days of the year. I guess we appreciate them more because they are so short. Winter bird watching is fun, and the birds are easier to see in the bare woods. The gray squirrel is left with its cache of stored nuts and the few red berries still hanging on the bushes. In the clear sky, you can see the star trails around Polaris, the North Star. You have to use a camera and take a time exposure to see them, though. By leaving the shutter open most of the night, you can get a picture of how the stars appear to move in circles around Polaris. But it's the earth that's rotating, so the camera is moving, not the stars.

Why is Polaris almost a steady point while the other stars seem to circle it? That's because the earth's axis is tilted right toward the North Star. If you poked a long stick

Winter is an excellent time for bird watching in Louisiana.

The earth's rotation makes the stars appear to move around the North Star.

The Big Dipper is high in the sky in the spring.

through the north and south poles and extended it up from the north pole, it would touch the North Star. Many ocean navigators use this and other stars to tell where they are in the ocean.

The Big Dipper is high in the sky in the spring. You can use it to help find the North Star. The stars in the Big Dipper are brighter than the North Star, and its noticeable shape is easy to find. The two stars at the end of the Dipper's cup are called the Pointers. They will lead you to the North Star. The North Star is also the tip of the handle to the Little Dipper. The skies have to be extremely clear for you to see this fainter constellation.

When the fresh green leaves burst out, the barren winter woods are soon forgotten. The gray squirrel perks up, waiting for the new buds, seeds, and nuts.

Knowing a little astronomy can help you find your way. The sun does not rise exactly in the east every day. In the winter, when the earth is tilted away from the sun, it rises toward the southeast, and in the summer toward the northeast. Only at the two equinoxes does it rise due east, at 90 degrees on your compass. But it crosses due south exactly halfway between sunrise and sunset every day of the year. So each day

Springtime's cool weather, like the fall's, makes hiking pleasant.

around lunchtime you can tell where the south is. If you're ever lost and need to know your direction, look at the noonday sun. North will be behind you, east to the left and west to the right. Then start hiking home.

The seasons show themselves in a host of ways. Summer is evident in the water of the swamp, fall in the vivid colors of the marsh, winter in the leafless trees along the bayou, spring in the fresh blooms on the vines.

Animal Behavior

The three primary needs that humans have are for food, shelter, and clothing. The rest of the animal world has the same primary needs, except that it has built-in clothing.

After those needs, an urge shared by all things in the wild kingdom is to continue their species by producing babies. Courtship and breeding behavior are some of the most interesting aspects of animal life.

Wading birds grow fancy plumes during their breeding season. They show these off during the rituals of courtship and nest building. Look for the beautiful plumes on the great egret each spring. Around cows you can see the cattle egret. In the spring and summer, some of the feathers on its back and face turn rusty red. Its beak also changes color in the breeding season.

Lubber grasshoppers mate.

A redwing blackbird sings.

Wild turkeys strut with their tail feathers spread, gobbling as they go. Theoretically, the biggest, strongest, fittest, and best gobbler is the one that will attract females to breed.

Redwing blackbird males sing from perches, proclaiming, This is my area. Those which have a good territory and can defend it well will draw a female.

Red wolves communicate by howling. One of my favorite sounds of the night is of wolves or coyotes sounding off.

Many animals, particularly insects, abandon eggs. But the fishing spider carefully carries her silk egg case

A turkey struts.

A red wolf howls.

A fishing spider carries her egg case.

around with her until its contents hatch.

Some birds practice deception to protect their young. A black skimmer waddles off, faking a broken wing to lead a predator away from her nest. When the predator catches up, she flies off, and the predator may have forgotten the nest.

A nutria dries and preens its fur every time it gets out of the water. Most mammals preen and wash themselves. Birds have to preen to keep their feathers in flying condition. Ducks have an oil gland that lets them preen with oil and stay waterproof. Insects preen. Fish rub themselves on logs and jump out of the water to knock parasites off. Without soap and shampoo, animals do a remarkably good job of keeping clean.

Make a list of other kinds of animal behavior you think of.

A black skimmer fakes injury.

A nutria preens.

Animal Homes

Blackbirds come in to roost.

A nutria lies in its den.

Animals need homes to raise their young, to hide from enemies, to avoid bad weather, and to store and at times to catch food. Some animals go about putting together their homes like architects, engineers, or even interior decorators; others are moochers, opportunists, or house thieves. Dirt, grass, twigs, branches, rocks, and mud are the construction materials that animals employ. Some make their own materials, such as paper, wax, or silk, for building their nests.

The sort of home an animal creates depends on the kind of habitat in which it lives and how much protection it needs. A Gulf Coast box turtle carries its home around with it but still burrows into a pile of vegetation during the winter, before it goes dormant. In the marshes and prairies, animals tunnel into

Two ducklings prepare to jump out of their nest box and follow their mother to water (bottom right). Nest boxes have helped increase the population of wood ducks. Other animals such as bees, raccoons, and opossums have benefited from such boxes, as well.

Some birds make their nests high over water to protect themselves from predators, as in this rookery of great blue herons, great egrets, and olivaceous cormorants.

the ground or make grassy nests. The nutria burrows into a muddy bank. In the forest, holes and cavities in trees are adopted as homes. Leaves and twigs go into the nests of many forest animals. Raccoons are among the mammals that prefer to settle into the cavities in old trees. Those homes are hard for raccoons to find these days, because many forests are cut down before the trees get old and hollow. Raccoons then seek out spots that strike them as similar, like the chimneys of our houses. A chimney sweep every so often gets a call to evict a coon. Because there are not enough tree cavities to shelter wood ducks, we put up boxes for them. House thieves, such as opossums, bees, raccoons, and owls, often move into these, but the wood ducks are benefiting from our effort. They're increasing in number despite the competition for their quarters.

Most animals come well dressed with fur, feathers, shells, or what have you, and they do not need a shelter all the time. The blackbirds that winter in Louisiana have roosts of millions. They have no roof over their heads, but they rest in the safety and company of great numbers.

When we think of nests, we think of trees. Herons nest high in trees for safety from predators. For the same reason, they also nest in groups. But if there are no land-based predators, birds can nest on the ground. On the small offshore islands of Louisiana, black skimmers, as well as terns, gulls, and pelicans, nest at or close to ground level.

Spider webs serve as homes as well as cupboards. Other insects that fly into the web are wrapped in silk and stored to be eaten later. Sometimes a spider spins itself a little room under a leaf. Do you think the leaf is protection from the rain or a screen to hide behind while it waits for prey?

Some homes are fairly simple, such as the tree bark under which a caterpillar lives.

Since we've done a lot to spoil the habitats of animals, we can do a little good by making birdhouses. Find out what kind of birds nest in your area, and put up a house in your backyard for that species.

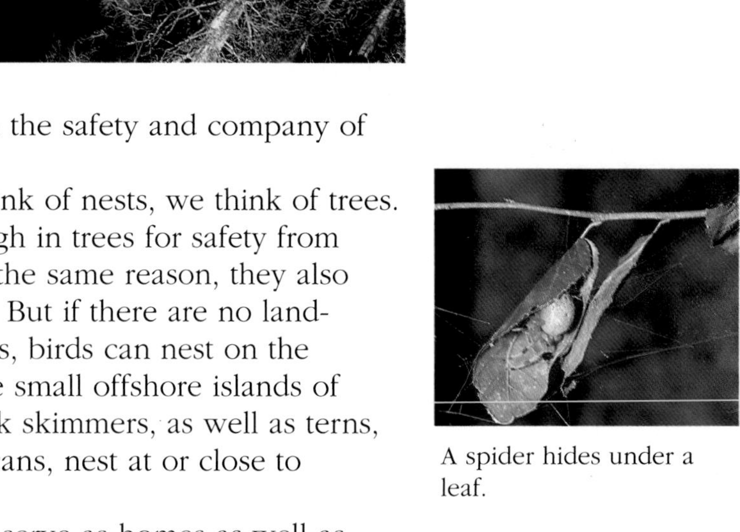

A spider hides under a leaf.

A caterpillar crawls under tree bark.

Finding Food

Grass, seeds, twigs, flowers, algae, nuts, tubers, insects, worms, fish, deer—this and much more is the food of animals. Every creature has to eat; feeding itself is its first and foremost need. Some animals hunt other animals; some graze or browse for vegetation. Some, like the turkey vulture, are part of nature's garbage disposal, for they eat dead animals.

Arthropods (insects, spiders, crawfish), because of their great numbers, end up in the diets of many animals. A great-crested flycatcher catches more than flies to feed its nestlings. A prothonotary warbler brings a fat, juicy caterpillar to its hungry young. It's total mayhem when an adult yellow-crowned night heron regurgitates a load of crawfish for its hungry babies.

Scavengers such as the turkey vulture scour the countryside of dead animals.

Fish Eaters

Louisiana has more water than most states, and in many forms: rivers, bayous, creeks, lakes, marshlands, and bays. There are plenty of fish in these waters. Naturally, we also have an abundance of fish-eating species. Animals have evolved with different ways of catching fish. Their body parts have developed in directions that make it easier for them to feed and survive.

A feeding frenzy erupts as five yellow-crowned night heron nestlings fight for the crawfish their mother has regurgitated.

Fish are constantly at risk, for many creatures—including other fish—want them for dinner. A bass engulfs a minnow (lower middle). An anhinga spears a fish (bottom), an osprey has dived for one (upper middle), and a water snake has swum for one (top).

The osprey has wings that let it hover and then dive for a fish. Sometimes completely underwater, it grabs the fish with strong, sharp talons. Back at its perch, it uses its knifelike beak to tear into the fish.

In contrast, a black skimmer's lower beak is longer than its upper, letting it knife through the waves, ready to close its mouth on any fish in its path. The great egret can stand patiently and grab a fish with its bill. The anhinga is the neatest of all, for it can swim under water and spear a fish there. It's comical to see the anhinga toss the pierced fish into the air, catching it headfirst in its beak to swallow it. Sometimes the fish is much larger than anything you'd imagine the bird could swallow.

A largemouth bass uses its split-second reflexes and suction like a vacuum cleaner's to swallow smaller fish. A diamondback water snake, like most snakes, can unhinge its jaws so it can open wide enough to swallow fish larger around than itself.

A bullfrog makes a crunchy snack for a bobcat.

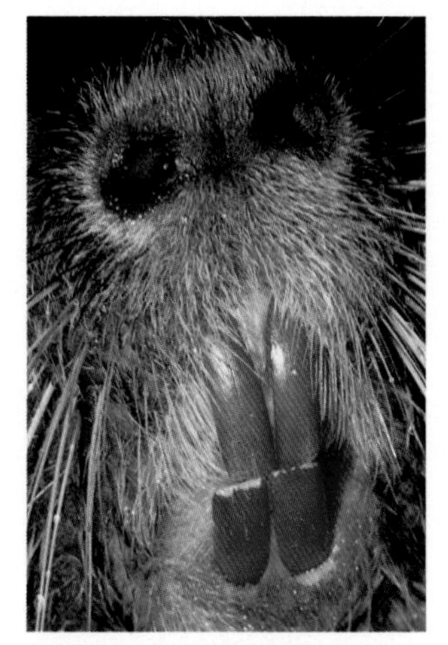

The nutria's incisors regrow as gnawing wears them down.

Plant Eaters

Human beings call themselves vegetarians when they eat only plant materials; they call animals on the same diet herbivores. The nutria is a herbivore that loves marsh grasses. Its orange-colored incisors are perfect for nibbling off pieces of the delicacy. Because the nutria's teeth get worn continuously, it, like most rodents, has teeth that keep growing throughout its life.

Two other rodents, the gray squirrel, seen here eating an acorn, and the flying squirrel, here eating a pecan, have teeth that renew themselves in the same way. Their chewing is best described as gnawing.

Many songbirds feed on seeds and nuts. The pine siskin likes little seeds suited to its little beak. The cardinal's conical bill can crack bigger and stronger seeds. But most seed-eating birds will feed their nestlings high-protein insects. The babies grow faster on an insect formula, and the rapidly reproducing fall webworms are stopped from

Nut and seed eaters include the gray squirrel (bottom left), the southern flying squirrel (bottom right), and the pine siskin (top right).

A fully grown young tern still begs food from its parents.

reaching the moon. Have you noticed how the beaks of baby birds so often look the same? They all have the same function: to open wide so the parents can stuff in lots of food. As the young mature, their beaks change to look just like their parents'.

Butterflies and hummingbirds like the nectar from flowers. In sipping all over the garden, they pollinate the plants.

Begging is the way most juvenile birds get by. Even though this young royal tern can fly, it still begs fish from its parents.

The Texas rat snake likes eating ro-

A Texas rat snake unhinges its jaw to swallow a wood-duck egg.

dents. It's a good snake to have on the farm. An excellent tree climber, it will eat eggs of wood ducks if it happens to luck into a nest of them. It too has a mouth that can open extra wide.

In the pyramid of life, it takes billions of seeds, fish, insects, and other lower forms to feed the big guys. Then the big guys get eaten too.

Watch the neighborhood birds. Notice their beaks, and try to guess what they eat.

A butterfly feeds on the nectar of a purple coneflower.

Locomotion

A cougar runs to catch its food.

A doe runs to escape danger.

Animals can walk, run, climb, fly, swim, slide, jump, crawl, or be carried. Each has evolved its own mode of locomotion.

A deer runs to escape cougars. A cougar runs to catch deer, but not quite so fast or so far as the deer. Occasionally a cougar using stealth as well as speed can catch a deer, but usually it can catch only the young and the crippled. Most land mammals walk or run without clumsiness. Some birds, such as the roadrunners and sandpipers, do as well.

The usual locomotion of birds is by flying. Owls can fly silently. Falcons can dive at 175 miles an hour. Some seabirds can stay in the air for months at a time. Each has survived because of the way it has evolved. Hawks, eagles, vultures, and gulls are expert at soaring. Many gulls, such as this laughing gull, will glide effortlessly behind a shrimp boat, patient for a tidbit to pop up from the net.

The long wings of a laughing gull are like those of a glider plane.

Coots need a runway (top left), but puddle ducks like teals can explode off the water (top right).

American coots have great paddles on their feet that they use not only in water but also for help in taking off into the air. Their chunky bodies and short wings don't let them explode off the water in the way a flock of blue-winged teals can.

Some animals have to depend on the wind to move them around, the way the Portuguese man-of-war jellyfish must.

Fish are excellent swimmers, as are most amphibians and reptiles. Observe how this cottonmouth swims on top of the water across a duckweed-covered bayou. Nonvenomous water snakes swim with their bodies under water and only their heads on the surface. Even a house cat can swim if it has to, but its close cousin the raccoon thrives in and around water. Many arthropods, such as some spiders, can walk on water.

By definition, plants can't move on their own from place to place, but many ways have evolved to move their seeds and pollen about. The cocklebur has fruit with tiny fishhook spines that catch onto animals' fur. Its seed is thus carried away to where the prickly balls drop off. Look closely at the spines. What everyday product do they look like? Do you think the inventor of Velcro got his idea here?

The cottonmouth's sinuous course in swimming is like its slither across the ground.

A cocklebur

Raccoons can run, climb, and swim very well.

Migration

Many animals migrate. Some migrate to nesting grounds, some to feeding grounds, and some to escape pests, weather, or enemies. We most often think of birds in connection with migration, but let's not forget about the butterflies, shrimp, and caribou.

One thing we have to remember in thinking about migration is that animals have no political boundaries like ours. There are 187 countries in the world, and each of these has its own customs—and its own game laws, if it has game laws at all. The enforcement of the hunting laws may be good in some countries, bad in others. Just think about the duck that has to survive the attitudes toward it of the governments of Canada, the United States, Mexico, and several Central and South American countries. Some of the countries do nothing to protect the waterfowl's habitat, and others have no game limits or do not enforce the ones they have.

On top of all the other reasons there are for getting along with the people of every country, there's the need to cooperate about migrating wildlife. Whether you say, "duck" (English), "pato" (Spanish), "anatra" (Italian), "ya" (Chinese), "canard" (French), or, "vit" (Vietnamese), you have to talk to others if you want to protect the duck.

Migration is awesome in what it involves. Just think of a tiny hummingbird crossing six

The Italian ancestors of Matthew Capprio said, "anatra."

The Spanish ancestors of Alan Nieves said, "pato."

Canadian geese, shrimp, butterflies, and ducks are among the animals that are international travelers, so no matter how you say, "duck," we need to work together to make the world livable for people and wildlife alike.

The British ancestors of Mary Carmichael said, "duck."

The Chinese ancestors of Ann Lin said, "ya."

hundred miles of the Gulf of Mexico. That's just a hop compared with the journeys of the arctic tern, which holds the record for distance of migration. It flies eleven thousand miles twice each year, between its breeding grounds in Greenland and its wintering grounds in the Antarctic.

To learn more about migration and to have one of the coolest bird-watching experiences possible anywhere, take a field trip to the Baton Rouge Audubon Society's Migratory Bird and Butterfly Sanctuary System (Holleman Nature Preserve), in Cameron Parish, between March and May. That's when all the songbirds return to their nesting grounds in Canada and the United States. Watch the weather, and time your trip for the passage of a cold front. The tiny birds will then have bucked a north wind all the way across the Gulf of Mexico and will be very tired. As soon as they see trees at the nature preserve, they'll just drop out of the sky. Ornithologists call this a fallout. If you're there for a good one, you'll get to see birds of every color: a vibrant scarlet tanager, a bright yellow prothonotary warbler, an electric blue indigo bunting. Grand Isle and all the other places with trees along the Louisiana coast are places to see this phenomenon, as well. The fall is another good time for coastal bird watching. Look for large groups of tree swal-

The French ancestors of Andre Savoie said, "canard."

The Vietnamese ancestors of Sarah Lambremont said, "vit."

This tired cattle egret stopped to rest on its northward migration across the Gulf of Mexico.

The scarlet tanager (middle) and prothonotary warbler (bottom) are two of the more colorful of the Gulf of Mexico's migrants. They land on coastal cheniers after their six-hundred-mile flights over open water.

lows. They mass and fly around like acrobats before starting across the Gulf.

Once while sailing from Mexico to New Orleans, I helped some birds migrate. A pair of barn swallows and cattle egrets alit on our boat to rest on their northward migration. They were welcome company.

Why do birds migrate? Scientists think it's to escape the winter and find food. What brings them back? We think they return because they're imprinted on their birthplace. How do they find their way? We speculate that they orient themselves by the stars, sun, and moon. Or maybe they detect the earth's magnetic fields and that helps them navigate. One thing we know for sure is that they don't use maps. Many questions remain unanswered, and ornithologists have lots left to study.

Keep a list of the different species of birds you see near your home during January, April, July, and October. You should notice a greater number of species during migration, in April and October.

Hiders and seekers both use camouflage. Humans use it, but it originated with animals. I use it in my nature photography, hunters use it to hide from game, and soldiers use it to hide in war. The LSU sweatshirt you might wear to a game is a sort of camouflage, since it makes you blend into the crowd. That's why animals evolved with camouflage: it lets them blend in, protect themselves from predators, and hide while hunting for food.

Like many human inventions, nature thought of it first. The sycamore tree can look just like a camouflage jacket.

Some animals, such as a green anole, can change colors from green to brown to match their surroundings as they hunt for insects or when they hide from sharp-eyed red-shouldered hawks. The anole is both predator and prey. Sometimes mistakenly called a chameleon, the anole is not related to those strange-looking creatures from the Old World.

The least bitterns stand erect in the reeds to camouflage themselves. Other birds, like the egrets and herons, which fish in shallow water on their branchlike legs, probably look like small bushes to fish swimming under water.

Mammals, both predator and prey, are masters of camouflage. Fawns have spots to

When the least bittern points its bill to the sky, it looks like the surrounding canes.

From a blind you can observe animals closely.

This zoo-raised white alligator would have died in the marsh, because it lacks normal camouflage colors.

break up their outline, and even the brown coats of adult deer can blend in with the forest. The camouflage is necessary for the fawn's survival, since predators like the bobcat, cougar, red wolf, and coyote would like to eat it.

Terns and skimmers nest on the beaches of our barrier islands. Both their eggs and their young blend in with the sand. When their parents are away, these skimmer chicks remain motionless and fully camouflaged.

Many sea creatures have chameleon-like abilities. Octopuses and flounder can change colors, and other ocean animals have appendages that look like the surrounding reef or seaweed.

Sometimes animals are born unpigmented. This condition is called albinism when the animals have pink eyes, leucism when their eyes are blue—like those of the famous white alligators at the Audubon Zoo, in New Orleans. If these gators weren't protected in the zoo, they would probably have been eaten when they were young, because they have no camouflage.

You can use camouflage yourself to observe animals, either by wearing it or by hiding in a blind. Try to observe backyard squirrels and birds from a blind. You can set up a tent or even just spread a blanket between chairs.

Using the natural brown pigment from ironstone and a camouflage paint kit, Drew Laxton has hidden himself in nature much as a green anole (top left), a white-tailed deer (top right), black skimmer chicks (bottom left), or a flounder (bottom right) would do. Perhaps the person who invented camouflage clothing got the ideas from the sycamore tree's bark (middle right).

Problems, Pollution, Solutions

Polluted air can come from automobile exhaust and factories, among other sources.

The Cameron Parish coast is eroding.

What humans do in a year or even a day can destroy a habitat that animals have spent millions of years adapting to. We must remember that we have to live on the earth too. To some, it may seem as if we can live despite the pollution of air and water and the misuse of pesticides. But even if we could, it would be a really sad life. Look how many people suffer from pollution-related diseases.

It's good that as a nation we are much more aware of our effect on other creatures' habitats today than we used to be. Most of us are trying to do things in a cleaner, better way, but there are so many of us and that in itself creates problems.

Coastal Erosion

Louisiana's coast is slowly melting away, and with it are going some of our most precious assets: our fish, shrimp, crabs, and more. Twenty-five square miles of land disappear from the Louisiana coast each year. This is productive marshland, important to our seafood industry. The erosion started when the levees went up along the Mississippi. They contained the muddy waters of the river that used to spread out over the low flat marsh and leave behind tons of sediment to refur-

Last Island, which once had a thriving resort hotel, is in danger of disappearing.

The straight man-made canal contrasts with meandering natural bayous. The canals bring salt water into the fresh marsh.

bish the coast. Now all that soil is pushed out into the Gulf of Mexico. So the marsh is sinking. We add to the problem by digging canals all over the coastal marsh. Salt water can move much faster in straight canals than in natural bayous with snakelike curves. Salt water kills the fresh-marsh plants, causing open water. Tidal movement also carries sediment out to sea. Boats make big waves that erode the canal banks. With less marshland, there's less room for our marine creatures to live and breed. That's having an effect on our number-one pastime, fishing, and on the satisfying of our number-one need, eating.

Pesticides

Pesticides and agricultural chemicals were developed to help farmers increase their crops. But stronger and stronger poisons were needed as the bugs evolved to resist some chemicals. Good bugs that ate bad bugs were killed too. A pattern unfolded. Farmers had to use more and more of stronger and stronger chemicals. Finally, fish, pelicans, and eagles were affected. Laws were passed. We banned some of the pesticides. But we're still putting tons of poisons in our fields and gardens every year. There are safer ways, cleaner ways, natural and organic ways, to take care of pests. Nature has handled population explosions for millions of years with animals. For example, bats and purple martins eat mosquitoes. Inventive and open minds can improve our life-styles without harming our environment. Remember that nature invented a lot of things first. Let's try to copy nature in pest control.

Clear-cutting is the main threat to our hardwood bottomlands.

Pesticides and urban runoff can cause fish kills.

The Recycling Cycle

The average American generates four pounds of trash a day and can dispose of ninety thousand pounds in a lifetime. The 261,875,000 people living in the United States today may throw away 23,568,750,000,000 pounds of trash before they die.

We have to recycle, or we'll end up in one big garbage pile. Most major towns in Louisiana have curbside recycling, or at least centers you can take recyclable materials to. Recycle. It is vital. But it's also very expensive. Each person needs to make less trash. Use things that are reusable—for example, glasses instead of plastic cups. Buy items with less packaging. Take your own canvas grocery bag to the market. There are a lot of ways to reduce what we throw away.

This snowy egret has no choice when it comes to our litter.

Illegal roadside dumps clutter the landscape.

The Clean Team helped get the first beach sweep started in 1985. With the help of environmental groups, state agencies, and private companies, tons of trash were picked up on beaches across Louisiana. Can you guess what item we picked up the most? Styrofoam coffee cups. If everyone carried reusable cups, not one bit of that litter would have been on our beaches. The Louisiana Pride Program has kept up the tradition of beach sweeps each fall. They've also started the Trash Bash to help clean the cities. Not only have these two festivals hauled heaps of litter away, they are also making people more conscious of trash. So a little less is being thrown down.

Drink cans are crushed before being recycled.

You can get involved in these enjoyable and worthwhile outings each year, and in the meantime, spread the word: Don't Trash Louisiana.

The Clean Team helps clear our beaches of the trash campers and fishermen leave, as well as that from oil rigs and passing ships (bottom left). You can help by joining the Clean Team each fall.

Endangered Species

Did you know that there could be over twenty billion bacteria in a quart of sour milk? Did you know that there are a hundred billion birds of various species on the earth? Can you guess how many insects there are? Most entomologists won't even speculate. Maybe millions of billions, and probably a lot more. One entomologist figures that if a single housefly bred successfully and lived for one year and if each of its offspring did the same, the entire earth would be covered with four feet of flies.

The human population is currently 5.5 billion. Not a lot compared with other animals, but look at the table to see how fast our numbers are growing now.

We reached our	**in the year**
first billion	1800
second billion	1927
third billion	1960
fourth billion	1975
fifth billion	1987
We'll reach our	**in the year**
sixth billion	1998
seventh billion	2008
eighth billion	2019
ninth billion	2032
tenth billion	2050

Will the sun set on our last sea turtle or will the growing human population learn to let both people and sea turtles swim freely in the sea of life?

Anthropologists estimate that agricultural human beings have been on the earth for ten thousand years. So let's do a little arithmetic. It took us 9,800 years to reach a world population of 1 billion human beings. To go from 1 billion to 2 billion took 127 years. But to go from 4 billion to 5 billion took only 12 years. That is accelerating growth. We humans aren't so numerous as insects or bacteria, but the 5.5 billion of us use and abuse the earth's resources like no other animal. Let's hope our brains are up to keeping the planet livable.

We aren't four feet deep in houseflies, because flies have a short life-span and a high percentage of their eggs, larvae, and adults are smashed, eaten, or killed by disease. Ever since the first green algae emerged on our planet, the survival game has been going on. Natural disasters and major climate changes have wiped out numerous species. We all know the story of the dinosaurs, but millions of other species have also died out because they couldn't compete. Today, though, human actions are the main cause of animals' endangerment.

Before 1973, when the United States Congress passed the Endangered Species Act, many of us didn't realize that there was a problem. We were changing natural habitats, poisoning animals through pollution, and overhunting them, without much awareness of the damage we were doing. We hunted the passenger pigeon into extinction. We hunted the buffalo into small herds in parks and zoos. We reduced the brown pelican, the peregrine falcon, and the bald eagle to very low numbers with the pesticide DDT. We destroyed the homes of the red wolf, the ivory-billed woodpecker, and the red-cockaded woodpecker by clearing bald cypress, bottomland hardwoods, and longleaf pines.

Today, with somewhat better pesticide controls and good wildlife management, we have brought the pelicans, eagles, and falcons back to where they can maintain their numbers. Leaving a few big trees in the forest has helped the red-cockaded woodpecker, but the passenger pigeon is extinct, the red wolf will probably never be able to reclaim its altered habitat from the coyote, and the ivory-billed woodpecker is gone from Louisiana, probably for good. We must hope that a few of these peckerwoods still live in Cuba.

A red wolf

Just remember that the human population is growing. As our numbers multiply, we have to be very careful how we use the earth.

Can we bring back the endangered sea turtles, or will we let the sun set on their skeletal remains?

The next time you go to the coast, count the brown pelicans you see. Remember that in 1962, there were none in Louisiana.

SOME OF LOUISIANA'S DEVASTATED SPECIES

Extinct
Passenger pigeon
Carolina parakeet
Eskimo curlew

Extinct in Louisiana
Attwaters prairie chicken
Whooping crane
Ivory-billed woodpecker

Endangered
Black bear
Red wolf
Florida panther
Brown pelican
Bald eagle
Peregrine falcon
Bachman's warbler
Gopher tortoise
Green sea turtle
Hawksbill sea turtle
Kemp's ridley sea turtle
Loggerhead sea turtle
Pallid sturgeon
Louisiana pearl shell
American burying beetle

A southern bald eagle

Backyard Wildlife

Almost anywhere you live, you can attract and view wildlife. You can have a bird feeder on your apartment's balcony in a big-city high-rise. But you'll see a greater variety of animals if you have a yard, especially if you make it pleasant for wildlife to visit. Your backyard can provide food, water, and cover. A bird feeder, a puddle or a dish of water, and a few shrubs for the animals to perch upon are the basics.

On the other hand, you can develop the landscape especially for wildlife. The National Wildlife Federation has a backyard habitat program that will offer suggestions about how to make the most of your grounds. It will recognize your yard with a certificate. For information, write to Backyard Wildlife Habitat Program, National Wildlife Federation, 1400 Sixteenth Street, N.W., Washington, DC 20036.

You can combine plants in such a way that there will be fruit and flowers in your yard throughout the year to attract and retain wildlife. Birdhouses help birds that need nesting cavities. Hedges and trees with

Backyard food (bottom) and water (top) will attract wildlife.

dense foliage attract nest builders. A small pond can attract frogs. I put crawfish in my pond and have seen egrets, herons, owls, and hawks fishing for them.

Feeders are favorites with birds and mammals. The rule for feeding wildlife, though, is not to make it totally dependent on you. You want the animals to use natural foods too, for that is how they'll get the full range of the minerals and vitamins they need. Dr. Gordon Pirie, the veterinarian at the Greater

Michael Matthews' patience lets him observe a black-chinned hummingbird at extremely close range.

Baton Rouge Zoo, has described how pet squirrels kept inside and fed only pecans have been brought to him with their bones broken in many places. A big juicy nut like the pecan is fine squirrel food, but squirrels also need sunshine and a diet of seeds, buds, bark, and leaves to keep their bones strong.

Raccoons are fun to watch under the backyard spotlight as they come up to nibble out of your dog's dish, but that can get out of hand. One family was putting out twenty-five pounds of dog food a night. I counted thirty-three raccoons the night I visited. The yard was a mess. It's no kindness to bring so many raccoons together to scarf on hand-outs. Diseases can spread, and the animals become too dependent on a single kind of food.

It's all right to catch and keep a toad, turtle, or garter snake for a few days. But let it go back to nature when you've finished observing it. Be sure to put it back where you found it.

Baby squirrels and raccoons that can't take care of themselves need expert care. If you run across these, you should turn them over to a rehabilitation center. Yet, sometimes baby animals are taken to such centers when they shouldn't be. That happens especially to fawns. When the doe senses an intruder,

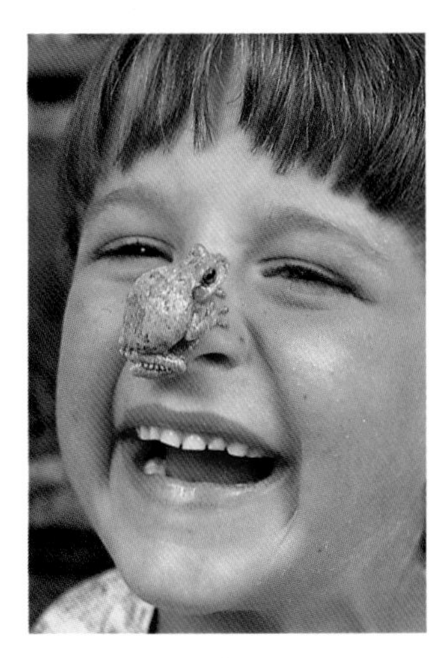

Jacob Roland clowns around with a gray tree frog he caught. As a good naturalist, he will release it where he found it.

By being too kind to backyard wildlife, you can do harm. When there are too many raccoons in a backyard, disease can spread.

Lee Gilly holds an orphan barred owl on his baseball glove. The owl has been under the care of a professional rehabilitator.

she runs off, leaving her spotted fawn camouflaged in the grass. She's relying on her fawn's not being seen, and she'll come back after the danger has passed. The mother deer is much better at taking care of a fawn than we'll ever be. Only if the mother is lying nearby, dead, should you take the fawn to a rehabilitation unit.

A baby bird that has fallen can be put back in its nest, and should be if possible. The mother won't abandon the nest. A fledgling sometimes jumps out a day early and hides itself in a bush. The mother usually comes to feed it and can do that much better than you can. If you end up with an orphaned bird, it's a lot of work to feed it every hour. Call your local center to get a proper diet for the species.

Visit a wild-animal rehabilitation unit. Many have programs for kids. It's interesting to learn how they train animals to go back to the wild.

Pineapple sage is a hummingbird favorite.

Trisha Screen observes a mourning dove incubating eggs on a friend's porch.

An important job in the protection of endangered species and their habitats is that of the wildlife manager who has trained as a biologist.

Tommy Edwards is a wildlife biologist for the Tensas National Wildlife Refuge, which was established to save some of the state's hardwood bottomland and protect the animals that make it their habitat. Tensas is one of the last places the ivory-billed woodpecker was seen in the United States. That was way back in 1943. It's also one of the last sanctuaries for the Louisiana black bear.

Tommy and Buck Marchinston, a graduate student, have trapped and radio-collared black bears both at the refuge and in the surrounding farmlands. Their goal is to learn more about what the bear requires in its habitat. They tell me that about sixty bears roam in Tensas and Madison parishes. The bears off the refuge live in small patches of forest between massive fields of soybeans and cotton.

The biologist uses snares. He and his assistant tranquilize any bear they catch, and they take its measurements and record its sex, estimated age, and whatever else they think matters. Then they put a radio collar on it that will allow them to track the animal by radiotelemetry. After giving the bear an antidote to the tranquilizer, the biologists leave. To follow the movements of the animal, they listen on headphones to beeps that a small directional antenna picks up. The louder the beep, the closer the bear.

From the data they collect, they can chart

A black bear drinks from a Louisiana swamp.

Wade Martin helps his dad track black bears near the Tensas National Wildlife Refuge (top). His dad, Tommy Edwards, who is a wildlife biologist at the refuge, sets a snare to catch a new bear (bottom).

the bear's movements on a map. They can find out what kind of place the bear selects for resting or for feeding. They learn where the bear goes as the seasons change. The idea is that by improving the Louisiana black bear's habitat, they'll help it make a comeback.

On one of my visits, we went to look for two tagged bears. For a couple of hours we walked, spying plenty of tracks and sometimes getting faint beeps. The big footprints of the bear were impressive. Tommy turned the earphones over to his stepson, Wade Martin, and we walked on. Wade was swinging the antenna around, trying to pick up the signal of Nora, an eight-year-old female. Tommy had tracked Nora for years and knew she should have cubs. He was also aware that she usually dens on the ground in a pile of brush, rather than, like most bears, in a hollow cypress tree.

Wade signaled us with a grin as he heard the beeps gets louder. We were on Nora's trail. We hunkered down and walked quietly until Tommy halted us and pointed to a black lump behind a brush pile. I set up my camera, and through the long lens and all the scraggly trees, I saw Nora with a cub in her mouth. She moved the cub a few feet away and went back for the others. We slipped away to leave her in peace. It was an experience I'll remember.

Another biologist who is a wildlife manager is Hillary Vincent, of the United States Fish and Wildlife Service. Her work is with another endangered species, the red-cockaded woodpecker. That seven-inch woodpecker has habitat needs different from any other species'. Old longleaf pine trees with red-heart disease are its only nesting sites. The diseased heartwood makes it easier for the woodpecker to drill out a nest hole.

Here Hillary and her helpers are banding the tiny featherless nestlings. It's a pretty scary job leaning on a ladder thirty feet up a pine tree trying to noose the babies out of their nest cavity. Once caught and banded, the birds are put safely back in the nest. Later the color-coded bands let the biologists know which bird they see and what nest it came from.

By learning more about the habitat needs of the red-cockaded woodpecker, wildlife managers will be able to adjust foresting practices to expand the habitat of that rare bird.

Hillary Vincent studies red-cockaded woodpeckers by banding week-old nestlings (left and bottom). A parent brings a grasshopper to its brood in a mature longleaf pine tree, the only place this species will make a nest (right).

Jobs in Nature

Leslie Tircuit coordinates the LSU Veterinary School raptor rehabilitation unit.

Occupations in the environmental field require greater numbers of workers with each passing year. By the time you grow up, three out of ten jobs will deal with nature, the environment, recycling, or waste management. So if you're interested in any of those fields, there should be openings for you.

When planning your career, stick to something you love. You'll enjoy life and perform better in your job if you like what you do. When your work is good, the money will follow. Look at the list on the next page of some of the many occupations that deal with nature and lend it a helping hand.

Jake Valentine, a biologist with the United States Fish and Wildlife Service, surveys colonies of terns.

Commercial fishermen, such as shrimpers, spend almost every day in the outdoors.

Dr. Paul Wagner leads students as well as adults on Honey Island swamp tours.

SOME OCCUPATIONS IN NATURE

Wildlife biologist
Marine biologist
Botanist
Ornithologist
Park ranger
Nature preserve manager
Zoo worker . . . many different jobs
Aquarium worker . . . many different jobs
United States Environmental Protection Agency employee . . . many different jobs
Louisiana Department of Environmental Quality employee . . . many different jobs
Natural history museum staff member
Nature center naturalist
Land manager
Nature-tour leader
Veterinarian
Landscape architect
Environmental lawyer
Environmental economist
Environmental lobbyist
Environmental organization staff member
Dealer in outdoor equipment and gear
Fishing guide
Hunting guide
Recycler
Artist
Writer
Wildlife photographer
Film maker
Teacher, professor
Farmer
Forester

Paul Yakupzack takes his daughter Amy J. around Cameron Prairie National Wildlife Refuge, which he manages.

Photographing Nature

A photography workshop in the Atchafalaya Basin teaches how a professional nature photographer works.

Be ready with your camera, and you may capture a comical moment.

Photographing, filming, or drawing nature is a great hobby or profession. I'm a lucky man, for I got a terrific job without planning for it. My high-school guidance counselor never mentioned a potential career in wildlife photography. He mentioned no nature-oriented jobs. But that was quite a while ago. Now there are plenty of work opportunities in nature. Photography, writing, and art are among them. Some practice in each of those areas will help you out in other careers as well.

You can learn a lot and enjoy yourself by visiting nature, art, or photography galleries. You get closeup views of animals and places you may never have a chance to see directly.

Nature books give the same kind of experience. Maybe you've been lucky enough to visit the Grand Canyon, but unless you move out to Arizona, you'll never know that place in winter, spring, summer, and fall unless you look at pictures. Good nature books help you get a feel for a place before and after you visit, or even if you don't visit at all.

Nature films are another easy way to come

in touch with wildlife and to learn about travel, geography, and photography. If you're interested in film and video, watch each scene closely and try to imagine the film maker's thinking as he spliced the story together.

Wildlife was probably the first subject matter of art, appearing many thousands of years ago in cave paintings. Today biologists make sketches in their field books, and professionals illustrate field guides and produce paintings, from abstract to realistic.

Many galleries, museums, nature centers, zoos, and environmental organizations schedule lectures and slide shows that can acquaint you with photography and art related to nature. Workshops in nature photography and art are available all over the world, and as close as in our own Atchafalaya Basin.

Here are some tips that can make your nature photographs better:

1. Learn to develop and print black-and-white film. Being able to see the whole process helps. If you get frustrated trying to make a good print from a bad negative, you'll take pains to have a better negative to work from next time.

2. Use a tripod whenever possible. Not only does a tripod steady your pictures but it commits you to taking extra time and thus gives you the opportunity to frame a better picture.

3. Take notes. Keep a record of the film, exposures, and weather. Then you'll know what you did on both your good pictures and your bad ones.

4. Go out into nature as often as you can. As long as you're visiting a swamp, a marsh, or a piny wood, there's a chance you'll see something new. You're certainly not going to have much of a chance sitting at home.

5. Practice, and always be ready. Whether you're photographing a sunset or a bird in flight, you can't be a slowpoke. Read your camera instructions thoroughly, practice holding your camera, and become very familiar with all its settings. With practice, you'll be ready to capture that decisive moment. Nature doesn't wait for you.

Terrie Frisbie is an artist who paints sand dunes in watercolors.

Tracks

Tracks spark curiosity in the budding naturalist. When you see a track in the mud or the sand, you wonder, What animal was it? What was it doing? Where was it going? Since most animals, especially mammals, are nocturnal or very secretive, tracks and other signs are usually all we observe. It's fun to discover what they can tell us.

The American Indians were expert trackers, since they depended on wild animals for food. Because we today do our hunting in supermarkets, tracking has become a lost art. But it's still fun to try to be an expert.

Two of the most frequently observed kinds of tracks in Louisiana are those left by raccoons and white-tailed deer. You can see these at the edge of just about any muddy stream or bayou. Turkeys, egrets, and herons leave fairly obvious tracks too.

To get a firsthand look at clear tracks, coax your dog or cat through some fresh mud and examine the footprints it leaves. Look at a field guide if you want to learn more about the differences between tracks. Sometimes you see not just footprints but also marks from tails, beaks, or wings. It

It's easiest to see tracks by a creek or in the mud. Raccoon tracks suggest that the animals congregated for food—maybe crawfish—the evening before (top). A boy compares his hand with a small deer's tracks in a sandy-bottomed creek (bottom).

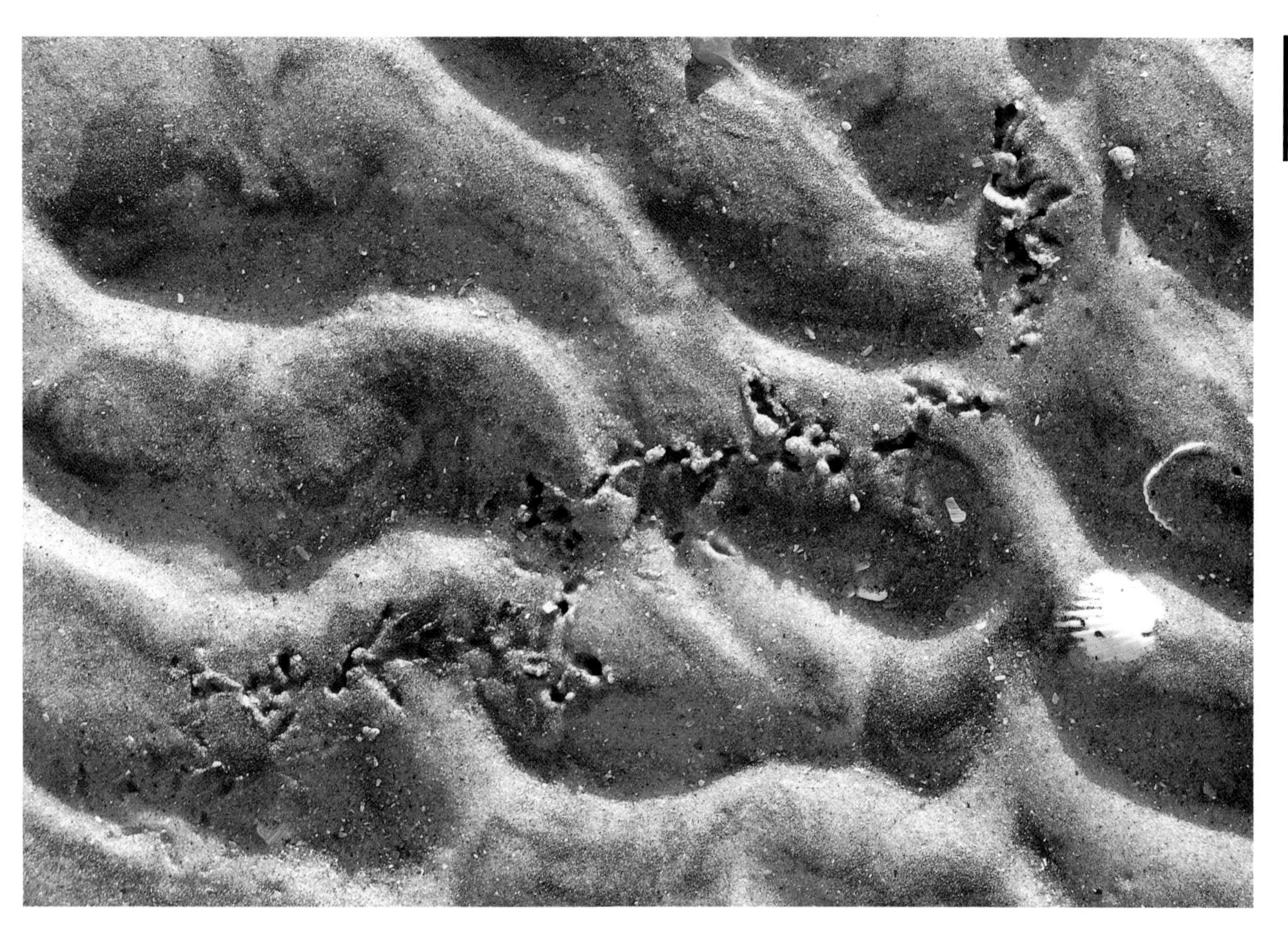

A sandpiper has left tracks with both its feet and its beak as it probed for invertebrates in the wet sand.

takes acute observation to become an expert tracker.

Humans, of course, are the easiest species to track, especially when they're thoughtless enough to leave a trail of litter. Leave your footprints in the woods, but take your litter home.

Dangerous Animals

An eastern diamondback rattlesnake.

How many deadly, dangerous kinds of things do you think there are in the Louisiana outdoors? Think of snakes, sharks, spiders, scorpions, snapping turtles, skunks, and sea urchins. And those are just the *s*'s. Aren't wasps, bees, barracudas, poison ivy, bull nettles, jellyfish, leeches, bears, coyotes, red wolves, and cougars dangerous too?

It's true that some of these things can be very dangerous, but none of them go out of their way to attack people. If you leave them alone, you'll be altogether safe.

Snakes are probably the most feared animals in the world. Some people seem born scared of snakes. Yet the snake, like any other animal, is a crucial part of our ecosystem. It's not just that it eats rodents. It's also part of the food supply for hawks, owls, and herons. People worry most about the venomous varieties, of which Louisiana has six. Five are pit vipers: a copperhead, a cottonmouth, and three species of rattlesnakes. In daylight, you can identify the pit vipers by the slitlike pupils of their eyes. The other is the coral snake, which is small and shaped like nonvenomous snakes, but very colorful. It has black, yellow, and red bands. When you see red bands touching yellow bands, you know you're looking at a coral snake. Here's a rhyme: Red on yellow, he's a nasty fellow.

Snakes don't attack people. But don't handle or get too close to a venomous snake. If you follow that rule, you won't get bitten.

The striped scorpion has a sting that hurts more than a yellow jacket's, but again, you'd have to be playing with it or be just unlucky for it to do anything to you.

A striped scorpion

A hornet's nest hangs at a second-story window.

A striped skunk

The spines of the bull nettle can be painful.

These creatures are secretive, living under tree bark or logs, and therefore they're rarely seen.

If you get close to a skunk, it will spray you, and boy will you ever stink. If you swim in the Gulf when a lot of jellyfish are around, you'll probably bump one and get stung. So don't swim with jellyfish. It's as simple as that. Common sense will protect you. If you pay attention to where you're going, you can avoid poison ivy, as well as the stinging spines of the bull nettle.

Animals are not vicious predators on human life. We have to get in their way to make them defend themselves. That's when some people get hurt. Stand back, enjoy observing, and they won't hurt you a bit.

The two exceptions to this rule that bother me in the Louisiana outdoors attack ferociously. These terrible creatures are gnats and mosquitoes. I slap them, and it's okay.

Try to think of a reason any animal would want to hurt you if you weren't bothering it.

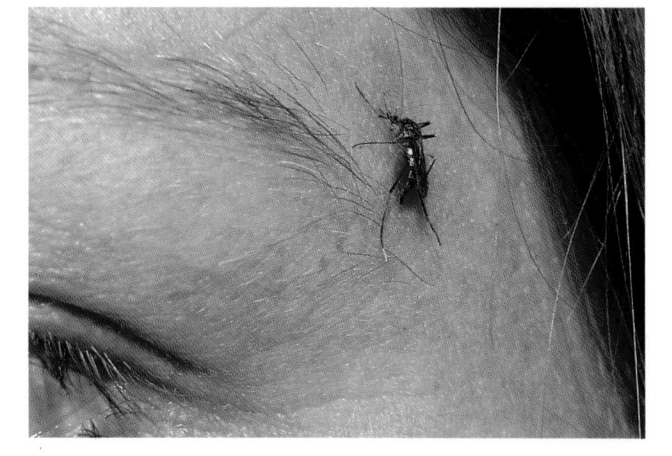

One of my very few enemies in nature is the mosquito.

State Animals, Plants, Minerals

Mammal	Black bear
Bird	Brown pelican
Dog	Catahoula leopard dog
Reptile	Alligator
Crustacean	Crawfish
Insect	Honeybee
Tree	Bald cypress
Flower	Magnolia
Wild flower	Louisiana iris
Fossil	Petrified palm wood
Gem	Agate
Amphibian	Green tree frog
Fish	Sac-à-lait

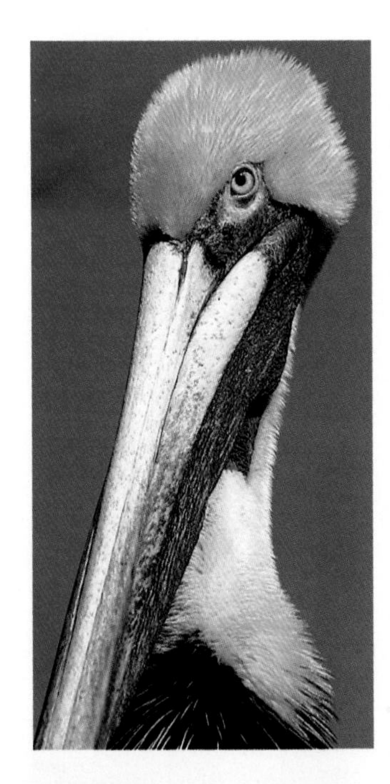

How to Take Your Parents Camping

It's not as hard as you think to take your parents camping. Dad may ask where, Mom may ask how, and both may ask why. Let's tackle the why first. One answer is, why not? Camping is fun.

You can also explain that camping relieves stress. Your parents' brains, ears, hearts, and souls need to get away from the phone, the television, and the city's smells and sounds. While camping in the piny woods of the Kisatchie National Forest, their senses and lungs will open to sensations the city has long ago shut them off from.

Being adorable may help win your folks over when nothing else will. Remind them that you're growing up fast. Time lost now can never be replaced.

Offer to teach them how to bird-watch or how to start a fire without matches or find huckleberries. If you don't know how, offer to learn with them. Be sure to tell the truth about the hardships of camping. You'll have to cook over a fire, and you may have to sleep through a rainstorm. Let's hope you'll be under a waterproof tent. Your parents may grumble, but just think of how satisfied they'll feel after roughing it successfully. They'll probably brim over with the gratitude they feel toward you.

Wayne Bettoney shows Robbie Welch the proper use of a compass. Even though the trail through the Castor Creek scenic area in Kisatchie National Forest is well marked, a map and a compass can help.

Where you take them depends on what they and you like to do best. There's good fishing in all four corners of the state, as well as swimming and canoeing. The legislature has designated a scenic river system of fifty-two rivers, streams, and bayous scattered around the state. For paddle trips, think of our small lakes, as well as our great swamps: the Atchafalaya and the Pearl River. The opportunities for hiking are a trifle more limited, especially for taking long hikes. The Kisatchie National Forest has a few long paths, with one of the nicest being the Wild Azalea Trail. The Tunica Hills offer both excellent hiking and rewarding fossil hunting.

Be sure to tell your parents to wear hunter's orange if you hike during deer season. Before you go, write to the park or refuge. See the list at the back of the book. Most of these places have maps and brochures that tell what they offer.

How to camp is something to be clear about before starting out. Unlike the mastodon-hunting Indians who had to camp every night in pursuing their food, we get to do it for fun. It's especially nice on nights when it's possible to sleep under the stars and not worry about rain or mosquitoes. Still, to enjoy yourself, you have to be prepared. Since I usually camp out about a hundred days a year, I've had time enough for this to be brought home to me.

First of all, you need your gear and supplies. Look at the checklist that follows this chapter. When you're car camping, you can bring all the stuff you want, but remember that some of the joy of camping comes from keeping things simple. Backpacking or canoe camping requires much more thought, planning, and packing. What you'll need depends on where you're going, for how long, and what you want to do when you get there.

The farther you go from the car, from a good road, and from a town, the more critical it is to have selected your emergency supplies intelligently. If you paddle far back in the swamp, you should have emergency food, water, matches, clothes, a spare paddle, and first-aid supplies just in case your trip takes a turn that you didn't expect.

Be sure you arrive early wherever you want to set up camp. If you get there two hours before dark, you'll be able to work comfortably in the daylight. Find a level spot for your tent but one that will drain well if it rains. Make sure no dead tree limbs hang overhead. Search for a place where you have a nice view, along with a breeze, in the summer, or a windbreak, in the winter. If there may be a lot of dew or rain, put your tent where it will get the morning sun. A wet tent is heavy to carry and will mildew.

When deciding what to take, choose unbreakables so far as you can. If you're going

Botanists join kids on a roller-coaster hike through the Tunica Hills Nature Preserve. The Nature Conservancy of Louisiana not only protects unique habitats in the state but also sponsors field trips to such places. Scientific experts conduct the visits.

to hike, pack light. Remove any wrapping that's unneeded. I have a friend who's a fanatic about lightening his load. He even drills holes in his toothbrush handle to get rid of a little weight. And remember, a good naturalist leaves the campsite cleaner than it was.

A Coleman or lightweight backpacker stove makes cooking easier, but I prefer the romance of cooking on an open fire. You can use a cast-iron skillet right on the fire only if you're car camping, because it's too heavy to carry on a hike. On backpacking trips, a small grill works well over two logs or two rocks. You can cook bread, chicken, or burgers on it. I like to mix together chopped potatoes, carrots, and onions, with a dab of butter and honey, to wrap in tinfoil and throw in the coals of the fire for a half hour. Marshmallows or s'mores (sandwiches made out of graham crackers, milk chocolate bars, and marshmallows) are a campfire tradition for dessert.

Fishing brings people into the Louisiana outdoors.

So that your parents can unwind and tune into the wilderness, get everybody at dusk to pick a direction and walk a hundred yards into the woods. After sitting down for a half hour in total quiet, listening to, smelling, and feeling the coming night, everyone can return to the campfire. Be sure each person has a flashlight to find the way through the dark. At the fire, share what you saw, heard, and felt. It's a good way to get close to nature. Then have your best storyteller relate a few ghost stories.

The crack of dawn is best for bird watching. Birds are active and hungry. In the springtime they are also establishing and defending territories. By making a "Pssh . . . Pssh . . . Pssh" sound with your teeth together and your lips opening and closing you can sometimes attract birds within ten feet of where you are.

Make sure your parents put out the fire completely before you leave. After all, you probably had to start it for them and keep it going all night. Make a last-minute check that you've gathered up all your belongings and litter.

If your parents are uncertain and think they don't know enough about camping to go, maybe with outside help, such as the Girl Scouts or the Boy Scouts provide, you'll be able to educate them. Those organizations hold group camp-outs, and the leaders can teach a lot. Shops that sell outdoor equipment, conservation organizations, and clubs also run camping field trips. It's not that hard, so let's take our parents camping.

Kids tumble in the rapids of Kisatchie Creek.

Bobby McCall displays techniques early settlers and American Indians used to build dwellings out of palmetto leaves. The kids are on a camp-out on Little Pecan Island.

Canoeing can take you places you probably wouldn't want to go if you had to swim. You can see sights like a massive bald-cypress stump (bottom left). The tree was already big when the Declaration of Independence was signed, in 1776.

Dancing flames form a backdrop for nighttime entertainment, after the day at Kisatchie Creek.

The best canoeing is at sunset.

Tammy Wood's fifth-grade class at the Park Ridge Elementary School, in Baton Rouge, nurtures a wetlands station, a small pond with aquatic plants. The class also maintains a nature trail through a forest in the school yard.

A cub-scout pack gives a wolf howl at a jamboree at Camp Avondale before an awards ceremony begins.

Kids get a closeup view of a nurse shark when volunteers bring New Orleans' Aquarium of the Americas to a grade-school classroom.

Zoos and aquariums let you see animals from other places. The sun bear is a favorite at the Audubon Zoo (top left), and the Caribbean reef tank draws viewers at the Aquarium of the Americas (bottom).

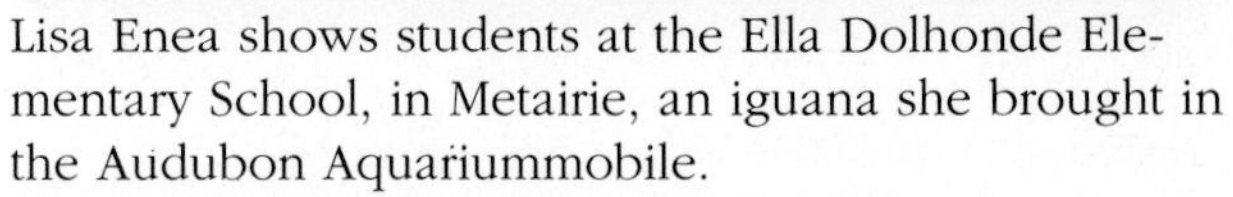

Lisa Enea shows students at the Ella Dolhonde Elementary School, in Metairie, an iguana she brought in the Audubon Aquariummobile.

Raphel Richard, a warden for the Nature Conservancy of Louisiana, shows Carrie Brantley and Martha Moore a baby alligator. Seeing animals close up is a great experience, but take proper care of the creatures and put them back in their natural habitat.

Pat Thomas scrutinizes a Florida cooter at the Louisiana Nature and Science Center, in New Orleans.

A special-education class in New Orleans is fascinated by an ostrich egg a guest lecturer hands around as part of a biology lesson.

Alaska Batchelor and Micah Luther get a closeup look at a fawn on National Hunting and Fishing Day. Each fall on this day, the Louisiana Department of Wildlife and Fisheries puts on lots of demonstrations for kids.

✓ Checklist for Field Trips

Essentials on Almost Any Outing

Matches
Compass
Maps
Sunscreen
Mosquito repellent
First-aid kit
Knife
Flashlight

Optional Gear

Camera
Tripod
Sketch pad
Journal
Fishing equipment
Field guides
Books

Gear for Day Hikes

Proper shoes for the terrain
Day pack
Water (two quarts per person on a long hot day)
Lunch
Trail snacks
Rain protection

Camping Gear

Backpack (if hiking)
Mosquito-proof tent
Ground cloth
Cookstove or small grill
Cooking and eating utensils
Biodegradable soap
Sleeping bag or blanket
Sleeping pad
Hatchet or machete
Campsite food
Snacks for trail
Emergency food
Spare clothes
Towel
Rope

Gear for Canoe Trips

Paddles, including a spare
Life jackets
Sponge and bailing cup
Boat cushion
Bow and stern rope
Tennis shoes (summer) or water-resistant boots (winter)
Waterproof bag for gear
Ice chest (can also protect gear)
Lunch
Snacks
Camping gear (if overnight)

Nature Places in Louisiana

National Wildlife Refuges

There is good wildlife viewing on the nature trails in some of the national wildlife refuges.

1. Bayou Cocodrie National Wildlife Refuge
   P.O. Box 1772
   Ferriday, LA 71334

2. Bayou Sauvage National Wildlife Refuge
   1010 Gause Boulevard, Building 936
   Slidell, LA 70458

3. Bogue Chitto National Wildlife Refuge
   1010 Gause Boulevard, Building 936
   Slidell, LA 70458

4. Cameron Prairie National Wildlife Refuge
   Route 1, Box 643
   Bell City, LA 70630

5. Catahoula National Wildlife Refuge
   P.O. Drawer Z
   Rhinehart, LA 71363-0201

6. D'Arbonne National Wildlife Refuge
   P.O. Box 3065
   Monroe, LA 71201

7. Lacassine National Wildlife Refuge
   HCR 63, Box 186
   Lake Arthur, LA 70549

8. Lake Ophelia National Wildlife Refuge
   P.O. Box 256
   Marksville, LA 71351

9. Sabine National Wildlife Refuge
   Highway 27, 3000 Main Street
   Hackberry, LA 70645

10. Tensas River National Wildlife Refuge
    Route 2, Box 295
    Tallulah, LA 71282

Kisatchie National Forest

The Kisatchie National Forest is divided into six districts spread out over north and central Louisiana. Good fishing, hiking, camping, hunting, and wildlife observation are available.

Kisatchie National Forest
2500 Shreveport Highway
Pineville, LA 71360

State Parks

Louisiana has twenty-seven state parks, preservation areas, and commemorative areas. Many have provisions for fishing, hiking, camping, and boating.

Office of State Parks
P.O. Box 94291
Baton Rouge, LA 70804

State Wildlife Management Areas

The state has forty-two wildlife management areas for public hunting and fishing. In some there are special regulations. The Department of Wildlife and Fisheries also administers five wildlife refuges, some of which are open to the public part of the year for fishing and wildlife viewing.

Department of Wildlife and Fisheries
P.O. Box 15570
Baton Rouge, LA 70895

Zoos, Aquariums, Wildlife Parks, and Arboretums

1. Alexandria Zoological Park
   P.O. Box 71
   Alexandria, LA 71309

2. Audubon Park and Zoological Gardens
   P.O. Box 4327
   6500 Magazine Street
   New Orleans, LA 70178
   M, CM, N, E, P, T, NP

3. Aquarium of the Americas
   One Canal Street
   New Orleans, LA 70130
   M, CM, N, E, P, T, NP

4. Briarwood
   Caroline Dormon Nature Preserve
   Saline, LA 71070
   M, T, NP

5. L. H. Cohn, Sr., Memorial Arboretum
   12056 Foster Road
   Baton Rouge, LA 70811

6. Global Wildlife Park
   Robert, LA 70455

7. Greater Baton Rouge Zoo
   3601 Thomas Road
   Baton Rouge, LA 70807
   M, E, P

8. Hilltop Arboretum
   11855 Highland Road
   Baton Rouge, LA 70808

9. W. B. Jacobs Memorial Nature Park
   80012 Blanchard Furrh Road
   Shreveport, LA 71107

10. Louisiana Purchase Gardens and Zoo
    Bernstein Drive
    Monroe, LA 71202

11. Louisiana State Arboretum
    Route 3, Box 168
    Shreveport, LA 71115

12. Zoo of Acadiana
    Highway 182
    Broussard, LA 70518

Natural History Museums

1. Acadiana Park Nature Station and Nature Trail
   East Alexander Street
   Lafayette, LA 70501

2. Bossier Parish Nature Study Center
   Linton Road
   Benton, LA 71006

3. Lafayette Natural History Museum
   116 Polk Street
   Lafayette, LA 70501

M = membership; CM = color magazine; N = newsletter; E = educational and environmental materials; P = meetings with programs; T = field trips; NP = nature preserves.

4. Louisiana Arts and Science Center
   100 River Road
   P.O. Box 3373
   Baton Rouge, LA 70821

5. Louisiana Nature and Science Center
   Joe Brown Park
   11000 Lake Forest Boulevard
   New Orleans, LA 70187-0610

6. Louisiana Tech Museum
   3089 Tech Station
   Ruston, LA 71272

7. Louisiana Wildlife and Fisheries Museum
   303 Williams Boulevard
   Kenner, LA 70062

8. Museum of Life Sciences
   8515 Youree Drive
   Shreveport, LA 71115

9. Museum of Natural Science, Louisiana State University
   Foster Hall
   Louisiana State University
   Baton Rouge, LA 70808-3216

10. Northeast Louisiana University Museum of Natural History
    3rd Floor, Hanna Hall
    Northeast Louisiana University
    Monroe, LA 71209

Private Swamp Tours

Numerous swamp and marsh tours are offered in south Louisiana. A complete list of those in a particular area is available from the tourist offices of the region. The Louisiana Nature and Science Center will also send you a list. I mention two tours that I have been on and recommend:

1. Annie Miller's Swamp Tours
   100 Alligator Lane
   Houma, LA 70360

2. Honey Island Swamp Tours
   106 Holly Ridge Drive
   Slidell, LA 70461

Nature Organizations

There are many groups, clubs, and organizations for people who wish to learn more about nature, protect the environment, or participate in field trips or classes.

1. National Audubon Society
   Membership Data Center
   P.O. Box 52529
   Boulder, CO 80322
   M, CM, E, NP, LC

   Acadiana Chapter, National Audubon Society
   c/o Mr. Joseph G. Vallee
   1006 West St. Mary Street
   Abbeville, LA 70510-3418
   M, N, P, T

   Baton Rouge Audubon Society
   P.O. Box 82525
   Baton Rouge, LA 70884-2525
   M, N, P, T, NP

   Central Louisiana Audubon Chapter
   c/o Mr. Patrick Ryan
   2008 Memorial Drive, Apt. 234
   Alexandria, LA 71301-3611
   M, N, P, T

   Natchitoches Audubon Society
   c/o Mr. Charles W. Harrington
   309 Watson Drive
   Natchitoches, LA 71457
   M, N, P, T

   Orleans Audubon Society
   P.O. Box 4162
   New Orleans, LA 70178-4162
   M, N, P, T

   Uncertain Audubon Society
   c/o Mr. Bill Wiener, Jr.
   401 Market Street, Suite 1110
   Shreveport, LA 71101-3270
   M, N, P, T

2. National Wildlife Federation
   1400 Sixteenth Street, N.W.
   Washington, DC 20036-2266
   M, CM, E, LC

   Louisiana Wildlife Federation
   P.O. Box 65239
   Baton Rouge, LA 70898
   M, N, E, LC

3. Sierra Club
   Member Services
   730 Polk Street
   San Francisco, CA 94109
   M, CM, E, LC

   Delta Chapter, Sierra Club
   P.O. Box 19469
   New Orleans, LA 70119
   M, N, P, T, LC

4. Nature Conservancy
   1815 North Lynn Street
   Arlington, VA 22209
   M, CM, E, NP

M = membership; CM = color magazine; N = newsletter; E = educational and environmental materials; P = meetings with programs; T = field trips; NP = nature preserves; LC = local chapters or affiliates.

Nature Conservancy of Louisiana
Box 4125
Baton Rouge, LA 70821
M, N, E, T, NP

5. Coalition to Restore Coastal Louisiana
8841 Highland Road, Suite C
Baton Rouge, LA 70808
M, N, E

6. Louisiana Ornithological Society
c/o Ms. Sammie Rodden
111 Lincoln Road
Monroe, LA 71203
M, N, E, T

For the addresses of other conservation organizations, visit your library. Among such organizations are the Wilderness Society, the Defenders of Wildlife, Ducks Unlimited, and the Isaac Walton League. For still other outdoor opportunities and education about the outdoors, consider the Boy Scouts of America, the Girl Scouts of America, the YMCA, and the YWCA, as well as shops selling outdoor and scuba diving equipment, local birding clubs, summer camps, and city and parish parks programs.

Index

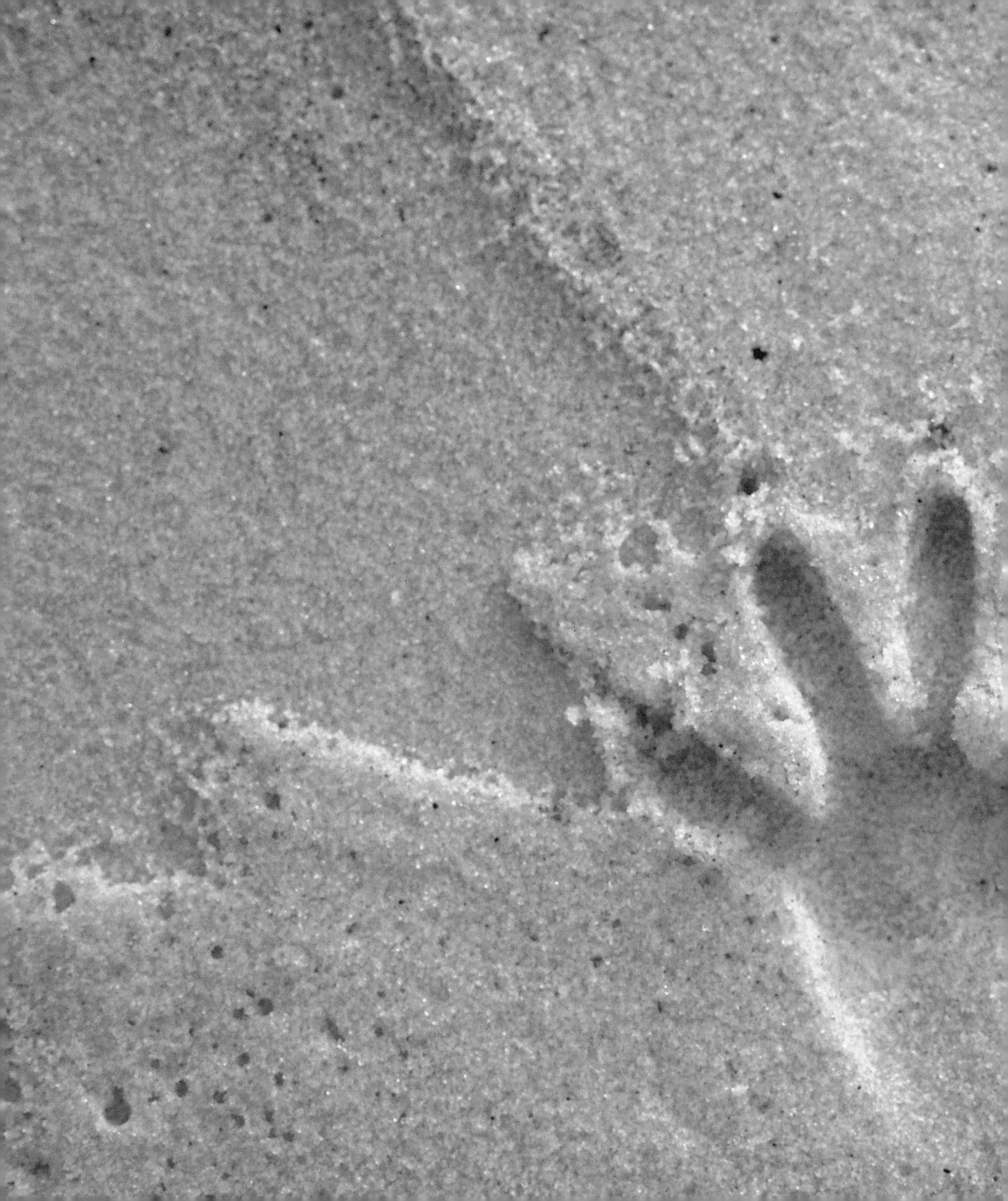